Basic Math Concepts

FOR WATER AND WASTEWATER PLANT OPERATORS

JOANNE KIRKPATRICK PRICE

Training Consultant

CRC Press
Taylor & Francis Group
Boca Raton London New York

CRC Press is an imprint of the
Taylor & Francis Group, an **informa** business

CRC Press
Taylor & Francis Group
6000 Broken Sound Parkway NW, Suite 300
Boca Raton, FL 33487-2742

© 1991 by Taylor & Francis Group, LLC
CRC Press is an imprint of Taylor & Francis Group, an Informa business

No claim to original U.S. Government works

ISBN-13: 978-0-87762-808-8 (hbk)

**Visit the Taylor & Francis Web site at
http://www.taylorandfrancis.com**

**and the CRC Press Web site at
http://www.crcpress.com**

Dedication

This book is dedicated to my family:

To my husband Benton C. Price who was patient and
supportive during the two years it took to write these texts, and
who not only had to carry extra responsibilities at home during
this time, but also, as a sanitary engineer, provided frequent
technical critique and suggestions.

To our children Lisa, Derek, Kimberly, and Corinne,
who so many times had to pitch in while I was busy writing,
and who frequently had to wait for my attention.

To my mother who has always been so encouraging and who
helped in so many ways throughout the writing process.

To my father, who passed away since the writing of the first
edition, but who, I know, would have had just as instrumental
a role in these books.

To the other members of my family, who have had to put up
with this and many other projects, but who maintain a sense
of humor about it.

Thank you for your love in allowing me to do something
that was important to me.

J.K.P.

Contents

Preface to the Second Edition

The first edition of these texts was written at the conclusion of three and a half years of instruction at Orange Coast College, Costa Mesa, California, for two different water and wastewater technology courses. The fundamental philosophy that governed the writing of these texts was that those who have difficulty in math often do not lack the ability for mathematical calculation, they merely have not learned, or have not been taught, the "language of math." The books, therefore, represent an attempt to bridge the gap between the reasoning processes and the language of math that exists for students who have difficulty in mathematics.

In the years since the first edition, I have continued to consider ways in which the texts could be improved. In this regard, I researched several topics including how people learn (learning styles, etc.), how the brain functions in storing and retrieving information, and the fundamentals of memory systems. Many of the changes incorporated in this second edition are a result of this research.

Two features of this second edition are of particular importance:

- the **skills check section** provided at the beginning of every basic math chapter

- a **grouping of similar types of calculations** in the applied math texts

The skills check feature of the basic math text enables the student to pinpoint the areas of math weakness, and thereby customizes the instruction to the needs of the individual student.

The first six chapters of each applied math text include calculations grouped by type of problem. These chapters have been included so that students could see the common thread in a variety of seemingly different calculations.

The changes incorporated in this second edition were field-tested during a three-year period in which I taught a water and wastewater mathematics course for Palomar Community College, San Marcos, California.

Written comments or suggestions regarding the improvement of any section of these texts or workbooks will be greatly appreciated by the author.

Joanne Kirkpatrick Price

Acknowledgments

"From the original planning of a book to its completion, the continued encouragement and support that the author receives is instrumental to the success of the book." This quote from the acknowledgments page of the first edition of these texts is even more true of the second edition.

First Edition

Those who assisted during the development of the first edition are: Walter S. Johnson and Benton C. Price, who reviewed both texts for content and made valuable suggestions for improvements; Silas Bruce, with whom the author team-taught for two and a half years, and who has a down-to-earth way of presenting wastewater concepts; Mariann Pape, Samuel R. Peterson and Robert B. Moore of Orange Coast College, Costa Mesa, California, and Jim Catania and Wayne Rodgers of the California State Water Resources Control Board, all of whom provided much needed support during the writing of the first edition.

The first edition was typed by Margaret Dionis, who completed the typing task with grace and style. Adele B. Reese, my mother, proofed both books from cover to cover and Robert V. Reese, my father, drew all diagrams (by hand) shown in both books.

Second Edition

The second edition was an even greater undertaking due to many additional calculations and because of the complex layout required. I would first like to acknowledge and thank Laurie Pilz, who did the computer work for all three texts and the two workbooks. Her skill, patience, and most of all perseverance has been instrumental in providing this new format for the texts. Her husband, Herb Pilz, helped in the original format design and he assisted frequently regarding questions of graphics design and computer software.

Those who provided technical review of various portions of the texts include Benton C. Price, Kenneth D. Kerri, Lynn Marshall, Wyatt Troxel and Mike Hoover. Their comments and suggestions are appreciated and have improved the current edition.

Many thanks also to the staff of the Fallbrook Sanitary District, Fallbrook, California, especially Virginia Grossman, Nancy Hector, Joyce Shand, Mike Page, and Weldon Platt for the numerous times questions were directed their way during the writing of these texts.

The staff of Technomic Publishing Company, Inc., also provided much advice and support during the writing of these texts. First, Melvyn Kohudic, President of Technomic Publishing Company, contacted me several times over the last few years, suggesting that the texts be revised. It was his gentle nudging that finally got the revision underway. Joseph Eckenrode helped work out some of the details in the initial stages and was a constant source of encouragement. Jeff Perini was copy editor for the texts. His keen attention to detail has been of great benefit to the final product. Leo Motter had the arduous task of final proof reading.

I wish to thank all my friends, but especially those in our Bible study group (Gene and Judy Rau, Floyd and Juanita Miller, Dick and Althea Birchall, and Mark and Penny Gray) and our neighbors, Herb and Laurie Pilz, who have all had to live with this project as it progressed slowly chapter by chapter, but who remained a source of strength and support when the project sometimes seemed overwhelming.

Lastly, the many students who have been in my classes or seminars over the years have had no small part in the final form these books have taken. The format and content of these texts is in response to their questions, problems, and successes over the years.

To all of these I extend my heartfelt thanks.

How To Use These Books

The *Mathematics for Water and Wastewater Treatment Plant Operators* series includes three texts and two workbooks:

- Basic Math Concepts for Water and Wastewater Plant Operators

- Applied Math for Water Plant Operators

- Workbook—Applied Math for Water Plant Operators

- Applied Math for Wastewater Plant Operators

- Workbook—Applied Math for Wastewater Plant Operators

Basic Math Concepts

All the basic math you will need to become adept in water and wastewater calculations has been included in the Basic Math Concepts text. This section has been expanded considerably from the basic math included in the first edition. For this reason, students are provided with more methods by which they may solve the problems.

Many people have weak areas in their math skills. It is therefore advisable to take the skills test at the beginning of each chapter in the basic math book to pinpoint areas that require review or study. If possible, it is best to resolve these weak areas before beginning either of the applied math texts. However, when this is not possible, the Basic Math Concepts text can be used as a reference resource for the applied math texts. For example, when making a calculation that includes tank volume, you may wish to refer to the basic math section on volumes.

Applied Math Texts and Workbooks

The applied math texts and workbooks are companion volumes. There is one set for water treatment plant operators and another for wastewater treatment plant operators. Each applied math text has two sections:

- Chapters 1 through 6 present various calculations **grouped by type of math problem**. Perhaps 70 percent of all water and wastewater calculations are represented by these six types. Chapter 7 groups various types of pumping problems into a single chapter. The calculations presented in these seven chapters are common to the water and wastewater fields and have therefore been included in both applied math texts.

  Since the calculations described in Chapters 1 through 6 represent the heart of water and wastewater treatment math, if possible, it is advisable that you master these general types of calculations before continuing with other calculations. Once completed, a review of these calculations in subsequent chapters will further strengthen your math skills.

- The remaining chapters in each applied math text include calculations **grouped by unit processes**. The calculations are presented in the order of the flow through a plant. Some of the calculations included in these chapters are not incorporated in Chapters 1 through 7, since they do not fall into any general problem-type grouping. These chapters are particularly suited for use in a classroom or seminar setting, where the math instruction must parallel unit process instruction.

The workbooks support the applied math texts section by section. They have also been vastly expanded in this edition so that the student can build strength in each type of calculation. A detailed answer key has been provided for all problems. The workbook pages have been perforated so that they may be used in a classroom setting as hand-in assignments. The pages have also been hole-punched so that the student may retain the pages in a notebook when they are returned.

The workbooks may be useful in preparing for a certification exam. However, because theses texts include both fundamental and advanced calculations, and because the requirements for each certification level vary somewhat from state to state, it is advisable that you <u>first determine the types of problems to be covered in your exam</u>, then focus on those types of calculations in these texts.

1 *Solving Math Problems*

SUMMARY

Those who experience difficulties with math are generally
facing two related problems:

- Difficulties with various types of math
  calculations, and, as a result,

- A distaste for math or anything closely related
  to it.

These two problems are very widespread. To verify this, ask
a few of your friends if they like math, or are good at it.
Even people who like math have areas in math where they
are weaker. Why is this?

DIFFICULTIES IN MATHEMATICS

There are at least four reasons people have had or continue to have difficulties in mathematics. Each of these problems can be addressed and resolved:

1. **A Poor Foundation.**

   Mathematics is primarily sequential—concept builds upon concept. If you miss a concept or are even weak in one of the concepts, calculations that build on that concept will suffer. It is a snowballing effect. People often conclude, "I can't do math".

   The solution to the poor foundation problem is simple (though not without effort): find your "weak spots" and resolve them. The skills check feature of this text is designed to help you find the specific problems that plague you. Once you have located the trouble spots, spend time on those topics until you have mastered them. The benefits will be well worth the effort.

2. **No Linking or Steps Missing.**

   When presented with a new concept, it is essential that the new concept be linked to something you already know. New ideas can become links in a "mental chain". Without that linking, remembering the concept is very difficult. A step-by-step presentation of material forges these links.

   The solution to the "no linking" problem is to be sure that any new material makes sense and connects with what you already know. Part of this linking occurs as you think about and ask questions about the new material. This is part of active learning. If you do not understand the new concept, ask others or consult other references. Find answers to your questions—this is part of your responsibility as a student.

3. **The "Big Picture" is Missing.**

   Understanding the overall concept, the "big picture", is integral to the learning process. It provides the skeleton on which all the details may be hung. Recall is improved because individual facts are part of a unified structure.

   The solution to the "big picture" problem rests largely in the hands of the textbook or instructor. Concerted effort has been made to include the underlying concepts for topics presented in this textbook and the companion textbooks.

4. **"Use It or Lose It" Syndrome.**

    "How easily or how long you remember something depends on how often you retrieve it. The more often you use it, the more accessible it becomes." This passage was written by Mort Herold in the book, *You Can Have A Near Perfect Memory*.

    This statement is also true in mathematics. The more you practice and use various math calculations, the easier they become.

    The solution to the "use it or lose it" syndrome is to use math calculations whenever possible. Rather than using basic math and applied calculations only when studying for or taking certification exams—incorporate as many calculations as possible in day-to-day operations. They will then become second nature to you.

SETTING UP AND SOLVING MATH PROBLEMS

There are basically two types of math:

- Theoretical math, and

- Applied math.

Theoretical math includes math concepts such as fractions, decimals, percents, areas, volumes, etc. These are the "tools" of math. The more tools you have at your disposal and the more adept you are at using them, the easier applied math problems will be.

Applied math is just what its name implies—basic math concepts applied in solving practical problems. The water and wastewater field is only one of many fields that utilize applied mathematics.

Those most successful in applied math calculations have a strategy, a way of approaching every problem that leads them methodically to the answer. Without such a strategy, students lose focus and many times get lost in the middle of the problem.

The strategies are simply the steps taken by the students, the questions they ask themselves, as they work their way through the problem. Using such strategies, students no longer have that helpless feeling—not knowing where to go or how to get there— a feeling so common to those who struggle with math.

A SUGGESTED STRATEGY

A basic strategy for solving math problems is described below. These steps can be used in solving any applied math problem.

1. **Disregarding all numbers, what type of problem is this?**

   For example, is this a conversion problem? A loading rate problem? A detention time problem? Recognizing the <u>type of problem</u> is the essential first step in solving it. Many times students will get so involved with numbers and units that they lose track of the basics of the problem.

   The first six chapters of the *Applied Math* texts have water and wastewater calculations grouped according to the type of math problem. This grouping should help you better identify and select the appropriate equation for the math problem.

2. **What diagram, if any, is associated with the concept identified in Step 1?**

   The heavy use of graphics throughout this math series is not accidental, nor is it intended primarily for aesthetic appeal.

   The inclusion of graphics whenever possible has several distinct advantages:

   - It helps maintain focus on the basic concept of the problem.

   - It helps organize data given in the problem and helps to determine additional data required, as the graphic is labeled with appropriate units.

   - It helps the student remember various aspects about that type of calculation. One of the strongest ways in which the brain sorts, stores, and associates facts is through visual images.

3. **What information is required to solve the problem and how is it expressed in the statement of the problem?**

   Many times the units given in the problem are not those required for the equation. Conversion of terms is an essential skill in solving the problems. If you cannot convert terms easily, you will be a prisoner of the memorized equation.

   Chapter 8 in this text describes conversion techniques and illustrates most of the conversions commonly required in water and wastewater calculations.

4. What is the final answer to the problem?

Once the equation has been selected, the diagram has been sketched, and units converted to desired terms, simply fill in the equation and solve for the unknown value.

It is suggested that you use one equation for each basic concept, and that you <u>do not rearrange</u> the equation for different unknown terms. For example, when using the $Q = AV$ equation, <u>use the same equation</u> , $Q = AV$, regardless of whether Q is the unknown, A is the unknown, or V is the unknown:

<u>Use:</u>

$$Q = AV$$

<u>Don't Use:</u>

$$A = \frac{Q}{V}$$

$$V = \frac{Q}{A}$$

The reason behind this suggestion has to do with how memory functions. Using the same equation over and over helps make the equation "second nature". Stated in terms of memory function, the more times you retrieve the same fact, the stronger the memory pathway to that fact. The terms of the equation may be rearranged, as needed, after the numbers have been placed in the equation. Chapter 2 describes the process of solving for the unknown value once the equation has been set up.

5. Does the answer make sense?

When you have obtained the answer to the calculation, be sure to ask yourself if the answer makes sense. Many times you will find a decimal error or some other obvious error at this point in the problem. The basics of estimation are described in Chapter 14 of this text.

Taking the time to think about the reasonableness of the answer is a key link in successful problem completion.

NOTES:

2 Solving for the Unknown Value

2.1 Solving for *x*—The Basics

Number
Correct

❏ Solve for *x* in each problem given below.

1. $8.1 = (3)(x)(1.5)$

$x =$ _____

6. $56.5 = \dfrac{3800}{(x)(8.34)}$

$x =$ _____

2. $(0.785)(0.33)(0.33)(x) = 0.49$

$x =$ _____

7. $114 = \dfrac{(230)(1.15)(8.34)}{(0.785)(70)(70)(x)}$

$x =$ _____

3. $\dfrac{233}{x} = 44$

$x =$ _____

8. $2 = \dfrac{x}{180}$

$x =$ _____

4. $940 = \dfrac{x}{(0.785)(90)(90)}$

$x =$ _____

9. $46 = \dfrac{(105)(x)(8.34)}{(0.785)(100)(100)(4)}$

$x =$ _____

5. $x = \dfrac{(165)(3)(8.34)}{0.5}$

$x =$ _____

10. $2.4 = \dfrac{(0.785)(5)(5)(4)(7.48)}{x}$

$x =$ _____

2.2 Solving for x^2

❑ Solve for x in the following problems.

1. $942 = (0.785)(x^2)(12)$

$x =$ _____

2. $6358.5 = (0.785)(x^2)$

$x =$ _____

3. $835 = \dfrac{4,200,000}{(0.785)(x^2)}$

$x =$ _____

4. $920 = \dfrac{3,312,000}{x^2}$

$x =$ _____

5. $23.9 = \dfrac{(3650)(3.95)(8.34)}{(0.785)(x^2)}$

$x =$ _____

2.3 Solving for x—Addition and Subtraction Problems

❑ Complete the following problems, solving for x as indicated.

1. $7 + 10 + x + 7 + 9 = 41$

$x =$ _____

2. $9.5 - x = 8.7$

$x =$ _____

3. $x + 93 = 165$

$x =$ _____

4. $10.1 = 9.5 + x$

$x =$ _____

5. $x + 15 = 19 + 22$

$x =$ _____

2.4 Solving for *x*—Advanced Problems

❑ Solve for the unknown value in the following problems.

1. $2x + 65 = 215$

$x =$ _____

2. $\dfrac{(4 + x)\,(2)\,(143)}{2} = 1320$

$x =$ _____

3. $0.042 = \dfrac{2085 + 876}{x + 29{,}190}$

$x =$ _____

4. $\dfrac{(6800)\,(8.34)\,(x) + (4500)\,(8.34)\,(0.07)}{56{,}712 + 37{,}530} = 0.051$

$x =$ _____

5. $7 = \dfrac{(2810)\,(0.435)\,(8.34)}{(6200)\,(x)\,(8.34) + (15)\,(1.8)\,(8.34)}$

$x =$ _____

NOTES:

2.1 SOLVING FOR *x*—THE BASICS

SUMMARY

1. When solving for the unknown variable (such as x), there are two basic objectives:

   - *x* must be in the numerator, and

   - *x* must be by itself (on one side of the equation).

2. To accomplish these objectives, only **diagonal movement of terms across the equal sign is permissible.**

In most water and wastewater treatment plant calculations, the unknown variable (*x* or any other variable) is part of a multiplication or division problem. The steps to use in solving these type problems are given in this section.

The problems selected for this section were chosen from actual problems presented throughout the water and wastewater texts.

MULTIPLICATION AND DIVISION PROBLEMS

When solving for the unknown variable, such as *x*, there are **two basic objectives:**

1. *x* must be in the numerator (top of the fraction), and

2. *x* must be by itself (on one side of the equation).

In solving equations, terms are moved from one side of an equation to the other with these two simple objectives in mind. **How the terms (numbers) are moved depends on the type of problem and how the numbers are related** For example, does the problems only involve multiplication and division of terms, or is addition or subtraction also indicated? Mathematical **rules of movement** and **order of operation** must be followed to obtain the correct answer to a calculation.

Most water and wastewater treatment calculations involve only multiplication and/or division of terms, with no addition or subtraction. For these basic calculations, only one rule of movement must be remembered: **move terms diagonally from one side of the equation to the other**.

Examples 1-3 illustrate solving for *x* using the diagonal movement.

ONLY ONE TYPE OF MOVEMENT IS PERMISSIBLE: DIAGONAL

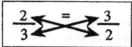

Make Diagonal Moves From One
Side of the Equation to the Other

Example 1: (Solving for *x*—The Basics)
❑ Solve for *x* in the equation given below.

$$(4 \text{ ft}) (1.5 \text{ ft}) (x \text{ fpm}) = 1100 \text{ cfm}$$

The two objectives in solving for x are: (1) *x* must be in the numerator; and (2) *x* must be by itself.

The first objective is satisfied since there is no number above *x*. (When no denominators are shown, a one is assumed to be the denominator of the fraction. Although it is not necessary, it may be helpful to rewrite the equation, as shown below.)

$$\frac{(4 \text{ ft}) (1.5 \text{ ft}) (x)}{1} = \frac{1100}{1}$$

To satisfy the second objective, the 4 and 1.5 must be moved away from *x* to the other side of the equation, in a diagonal move (from the top of one side of the equation to the bottom of the other side).

$$x = \frac{1100}{(4.1) (1.5)}$$

The value of *x* can now be calculated:

$$x = \boxed{179 \text{ fpm}}$$

Example 2: (Solving for *x*—The Basics)
❑ Given the following equation, solve for the unknown *x*.

$$730 = \frac{x}{3847}$$

x is in the numerator, so the first objective has been satisfied. Now clear for *x* (leave *x* where it is and move all other numbers from the "*x*-side" of the equation to the other side):

$$(\quad\longleftarrow)\,(730) = \frac{x}{3847}$$

The unknown value (*x*) can now be calculated:

$$(3847)\,(730) = x$$

$$\boxed{2{,}808{,}310} = x$$

Example 3: (Solving for *x*—The Basics)
❑ Given the equation shown below, solve for the unknown *x*.

$$0.5 = \frac{(165)\,(3)\,(8.34)}{(x)}$$

To satisfy the first objective, *x* must be moved from the denominator of one side of the equation diagonally to the numerator of the other side:

$$\longleftarrow 0.5 = \frac{(165)\,(3)\,(8.34)}{(x)}$$

$$(x)\,(0.5) = (165)\,(3)\,(8.34)$$

Now clear for *x* (move all other terms away from *x*)

$$x = \frac{(165)\,(3)\,(8.34)}{0.5}$$

$$x = \boxed{8257}$$

WHY IS DIAGONAL MOVEMENT OF TERMS REQUIRED?

Diagonal movement of terms from one side of an equation to the other is necessary so that the value of the equation is not changed.

From a practical standpoint, if you move the terms any other way than in a diagonal, you are changing or distorting the value of the original equation. To understand why this is so, we will need to consider more closely the math concepts associated with moving the terms.

First of all, an equation is a mathematical statement in which the terms or calculation on one side equals the terms or calculation on the other side. **To preserve this equality, anything done to one side of the equation must be done to the other side as well.** For example, suppose you must solve the equation:

$$3x = 14$$

To solve for *x*, you will want to get *x* by itself. Since *x* is multiplied by 3, you can get rid of the 3 by using the opposite process—dividing by 3. However to preserve the equation, you must divide the other side of the equation as well:

$$\frac{\cancel{3}x}{\cancel{3}} = \frac{14}{3}$$

$$x = \frac{14}{3}$$

Since both sides of the equation are divided by the same number, the value of the overall equation remains unchanged. **Notice that from a practical standpoint, the 3 was moved from the top of the left side to the bottom of the right side of the equation —a diagonal move.**

PRACTICE PROBLEMS 2.1: Solving for *x* —The Basics

❏ Solve for *x* in the problems given below.

1. $19{,}747 = (20)\,(12)\,(x)\,(7.48)$

$x =$ _____

2. $\dfrac{(15)\,(12)\,(1.25)\,(7.48)}{x} = 337$

$x =$ _____

3. $\dfrac{x}{(4.5)\,(8.34)} = 213$

$x =$ _____

4. $\dfrac{x}{246} = 2.4$

$x =$ _____

5. $6 = \dfrac{(x)\,(0.18)\,(8.34)}{(65)\,(1.3)\,(8.34)}$

$x =$ _____

6. $\dfrac{(3000)\,(3.6)\,(8.34)}{(0.785)\,(x)} = 23.4$

$x =$ _____

7. $109 = \dfrac{x}{(0.785)\,(80)\,(80)}$

$x =$ _____

8. $(x)\,(3.7)\,(8.34) = 3620$

$x =$ _____

9. $2.5 = \dfrac{1{,}270{,}000}{x}$

$x =$ _____

10. $0.59 = \dfrac{(170)\,(2.42)\,(8.34)}{(1980)\,(x)\,(8.34)}$

$x =$ _____

2.2 SOLVING FOR x^2

SUMMARY
1. When solving for x^2, first follow the same procedure as solving for x:
• Verify that x^2 is in the numerator. If it is not, move terms according to the diagonal rule.
• Verify that x^2 is by itself. If it is not, move terms away from x^2 according to the diagonal rule.
2. Once x^2 is in the numerator and by itself:
• Complete the calculation indicated on the opposite side of the equation from x^2.
• Then take the square root of both sides of the equation.

Occasionally the unknown variable will be a squared term, such as x^2. When this is the case, follow all the steps in solving for x, as described in the previous section, then use a calculator to take the square root.

In calculations which involve area or volume, sometimes the unknown value will be a squared term, such as x^2. For example, the cross-sectional area of a pipe may be known but not the diameter of the pipe. Or the area of a square may be known, but the length of the side unknown. In calculations such as these, the unknown value is a squared term (x^2 or D^2, etc).

To solve problems involving squared terms:

• First, **follow the same procedures as solving for x.** (Move terms as necessary so that x is in the numerator, and x is by itself.)

• Complete the calculation indicated on the opposite side of the equation from x^2.

• Then **take the square root* of both sides of the equation.**

Examples 1-3 illustrate how to solve for a squared term.

ONCE x^2 IS IN THE NUMERATOR AND BY ITSELF, TAKE THE SQUARE ROOT OF BOTH SIDES OF THE EQUATION

$$x^2 = 15{,}625 \text{ ft}^2$$

$$\sqrt{x^2} = \sqrt{15{,}625 \text{ ft}^2}$$

The square root of x^2 is always x

Use a calculator to determine the square root. Enter 15,625 then press the square root function ($\sqrt{}$)

$$x = \boxed{125 \text{ ft}}$$

Example 1: (Solving for x^2)
❑ Given the problem below, find the value of x.

$$(0.785)\,(x^2) = 2826$$

First, move terms so that x^2 is in the numerator and by itself: (Remember, only diagonal moves are permitted.)

$$x^2 = \frac{2826}{0.785}$$

Simplify the fraction before taking the square root:

$$x^2 = 3600$$

Now take the square root of both sides of the equation:

$$x = \boxed{60}$$

---

* For a discussion of square roots, refer to Chapter 13.

Example 2: (Solving for x^2)
❏ Find the value of D in the problem given below.

$$70.8 \; = \; \frac{50,000}{(0.785) \, (D^2)}$$

First, move terms using the diagonal rule, so that D^2 is in the numerator and by itself: (D^2 and 70.8 can "switch places", leaving D^2 in the numerator and by itself.)

$$D^2 \; = \; \frac{50,000}{(0.785) \, (70.8)}$$

Next, complete the calculation indicated on the right side of the equation:

$$D^2 \; = \; 900$$

Then take the square root of both sides of the equation:

$$D \; = \; \boxed{30}$$

Example 3: (Solving for x^2)
❏ What is the diameter, D, of the tank given the following equation?

$$1.5 \; = \; \frac{(0.785) \, (D^2) \, (12) \, (7.48)}{198,000}$$

First, move terms so that D^2 is in the numerator and by itself: (Since D^2 is already in the numerator, leave D^2 on the right side of the equation and move all other terms to the left side of the equation. Remember to move terms diagonally.)

$$\frac{(1.5) \, (198,000)}{(0.785) \, (12) \, (7.48)} \; = \; D^2$$

Complete the calculation indicated:

$$4215 \; = \; D^2$$

Then take the square root of both sides of the equation:

$$\boxed{65} \; = \; D$$

PRACTICE PROBLEMS 2.2: Solving for x^2

1. $(0.785)(D^2) = 5024$

$D = \underline{\hspace{2cm}}$

2. $(x^2)(10)(7.48) = 10{,}771.2$

$x = \underline{\hspace{2cm}}$

3. $51 = \dfrac{64{,}000}{(0.785)(D^2)}$

$D = \underline{\hspace{2cm}}$

4. $(0.785)(D^2) = 0.54$

$D = \underline{\hspace{2cm}}$

5. $2.1 = \dfrac{(0.785)(D^2)(15)(7.48)}{(0.785)(80)(80)}$

$D = \underline{\hspace{2cm}}$

2.3 SOLVING FOR *x*—ADDITION AND SUBTRACTION PROBLEMS

SUMMARY

1. When solving for the unknown variable (such as *x*) involving addition and subtraction, there are two principal objectives:

   - *x* must be positive, and

   - *x* must be by itself (on one side of the equation).

2. To accomplish these objectives, the terms may be moved from one side of the equation to the other by **changing the sign of the term moved** (from positive to negative, or vice versa).

$$7.8 = x + 0.4$$

   In order to get *x* by itself, 0.4 must be moved to the left side of the equation. The sign in front of 0.4 changes from positive to negative.

$$7.8 - 0.4 = x$$

When calculations involve addition or subtraction, diagonal movement of the terms is not permissible. The rules of movement are summarized to the left.

The calculations described in this section are limited to those that involve addition and subtraction only. Calculations which incorporate multiplication and division as well as addition and subtraction are discussed in Section 2.4.

ADDITION AND SUBTRACTION PROBLEMS

Two water and wastewater calculations that involve only addition and subtraction are perimeter problems and chlorine dosage/demand/residual problems.

As in other problems when solving for x, the basic objective is the same:

- Get x by itself on one side of the equation.

In order to do this, terms must often be moved from the "x-side" of the equation to the other side of the equation. When moving an addition or subtraction term, change the sign of that term. Examples 1-3 illustrate this type of calculation.

WHEN MOVING ADDITION OR SUBTRACTION TERMS, CHANGE THE SIGN OF THE TERM

$$45 + x + 62 + 76 = 237$$

$$x = 237 - 45 - 62 - 76$$

$$x = \boxed{54}$$

Example 1: (Solving for x—Addit. & Subtr.)
❑ Given the following equation, solve for x:

$$115 + 105 + 80 + x = 386$$

x is positive, so the only other criteria to be met is that x must be by itself. Therefore, the three numbers 115, 105, and 80 must be moved away from x to the right side of the equation: (Remember to change the signs from positive to negative.)

$$x = 386 - 115 - 105 - 80$$

$$x = \boxed{86}$$

Example 2: (Solving for *x*—Addit. & Subtr.)
❑ Solve for *x* in the problem shown below:

$$6.3 - x = 5.5$$

x must first be positive in order to solve for *x*. Therefore *x* must be moved to the other side of the equation so that the sign may be changed to positive:

$$6.3 = 5.5 + x$$

Now get *x* by itself by moving 5.5 away from *x* to the other side of the equation:

$$6.3 - 5.5 = x$$

$$\boxed{0.8} = x$$

Example 3: (Solving for *x*—Addit. & Subtr.)
❑ Find the value for *x* in the equation shown below.

$$17 + 23 + 7 - x = 38$$

First, move *x* to the other side of the equation so that it will be positive:

$$17 + 23 + 7 = 38 + x$$

Then get *x* by itself. In this example, 38 must be moved to the other side of the equation:

$$17 + 23 + 7 - 38 = x$$

$$\boxed{9} = x$$

WHY IS A CHANGE OF SIGNS REQUIRED?

It is necessary to change the sign of an addition or subtraction term when it is moved from one side of an equation to the other. Why is this?

The mathematical reason for this is similar to that described in Section 2.1 regarding diagonal movement of multiplication or division terms. That is, since an equation has equal terms or calculations on each side of the equation, anything done to one side of an equation must be done to the other side of the equation as well.

For example, suppose you wish to solve the equation

$$42 + x + 19 = 83$$

Since *x* must be by itself, 42 and 19 must be moved to the right side of the equation. Both of these terms are positive. (Whenever a number does not have a sign indicated, such as 42, it is assumed to be positive.) Therefore to move these two terms from the left side of the equation, the opposite process (subtraction, or negative numbers) must be used. And to preserve the equation, if you subtract a number from one side of an equation, you must also subtract it from the other side:

$$42 + x + 19 \boxed{-42 - 19} = 83 \boxed{-42 - 19}$$

$$x = 83 - 42 - 19$$

$$x = \boxed{22}$$

Note that from a practical standpoint, the positive numbers on the left side of the equation were changed to negative numbers when moved to the right side of the equation—**a change from positive to negative.**

PRACTICE PROBLEMS 2.3: Addition and Subtraction Problems

❏ Find the value of x in the problems given below.

1. $6.7 = x + 0.9$

4. $19 + x - 14 = 16 + 3$

$x =$ _____

$x =$ _____

2. $16 + x + 18 + 42 = 121$

5. $82 + 50 - x = 109$

$x =$ _____

$x =$ _____

3. $9.8 - x = 7.9$

$x =$ _____

2.4 SOLVING FOR *x*—ADVANCED PROBLEMS

SUMMARY

To solve for *x* when multiplication and division as well as addition and subtraction of terms is indicated, use the following steps:

1. Simplify as many terms as possible, using the order of operation:

   - Complete all multiplication and division from left to right.

   - Complete all addition and subtraction from left to right.

2. Verify that the *x* term is in the numerator. If it is not, move the *x* term to the numerator, using a diagonal move.

3. Verify that *x* is by itself, on one side of the equation.

In the more advanced water and wastewater calculations, you will sometimes encounter problems that involve multiplication and division as well as addition and subtraction. To solve for *x* in these problems, the specific order of operation must be followed.

Some of the more complex water and wastewater calculations include addition and subtraction of terms as well as the usual multiplication and division. Solving for x in these calculations involves a couple of considerations not found in the basic problems.

The three basic steps in solving for x in more advanced calculations may be summarized as shown to the right.

Let's examine each of these three steps more closely. In **Step 1**, you are to simplify as many terms as possible using the order of operation.

Sometimes parentheses or brackets (called grouping symbols) contain multiplication and division problems, etc., such as:

$$\frac{4\,(10+3)+6}{2}$$

When this is the case, **simplify within the grouping symbols first**, using the order of operation; then continue with the rest of the calculation, using the same order of operation. For example, in the calculation above, the steps to simplify would be as follows:

$$\frac{4\,(10+3)+6}{2}$$

$$= 4\,(5+3)+6$$
$$= 4\,(8)+6$$
$$= 32+6$$
$$= 38$$

THREE STEPS IN SOLVING FOR X IN ADVANCED CALCULATIONS

1. Simplify as many terms as possible, using the mathematics **order of operation**:

   - Complete all multiplication and division from left to right, and

   - Complete all addition and subtraction from left to right.

2. Verify that the x term is in the numerator. If it is not, move it diagonally across the equal sign to the numerator on the other side of the equation.

3. Get x by itself on one side of the equation.

Example 1: (Solving for x—Advanced)
❑ Solve for x in the problem given below:

$$\frac{(10+x)\,(4)\,(120)}{2} = 3840$$

First, simplify as many terms as possible:

$$\frac{(10+x)\,(480)}{2} = 3840$$

Since x is already in the numerator, now get x by itself. (10 is connected to x by an addition sign. It is therefore temporarily considered part of the x term). Move the 2 and 480 away from the x term, using the diagonal rule:

$$10+x = \frac{(3840)\,(2)}{480}$$

The right side of the equation can be simplified:

$$10+x = 16$$

Now complete the problem by getting x by itself:

$$x = 16 - 10$$

$$x = \boxed{6}$$

Example 2: (Solving for x—Advanced)
❑ Find the value of the problem given below:

$$0.049 = \frac{x + 1020}{18,200 + 21,400}$$

First, simplify as many terms as possible:

$$0.049 = \frac{x + 1020}{39,600}$$

x is in the numerator so now get the x term by itself.
(x + 1020 are connected by a plus sign and are temporarily considered the x term.)

$$(0.049)(39,600) = x + 1020$$
$$1940.4 = x + 1020$$
$$1940.4 - 1020 = x$$
$$\boxed{920.4} = x$$

Example 3: (Solving for x—Advanced)
❑ Solve for x in the equation given below.

$$\frac{(2780)(0.545)(8.34)}{(6200)(x)(8.34) + (16)(1.8)(8.34)} = 7.2$$

First, simplify as many terms as possible:

$$\frac{12,636}{(6200)(x)(8.34) + 240} = 7.2$$

x must be in the numerator. Since the terms in the denominator are connected by a plus sign, the entire string of numbers is moved up into the numerator of the opposite side:

$$12,636 = (7.2)\left[(6200)(x)(8.34) + 240\right]$$
$$\frac{12,636}{7.2} = (6200)(x)(8.34) + 240$$
$$1755 = (6200)(x)(8.34) + 240$$
$$1755 - 240 = (6200)(x)(8.34)$$
$$1515 = (6200)(x)(8.34)$$
$$\frac{1515}{(6200)(8.34)} = x$$
$$\boxed{0.029} = x$$

In **Step 2** of solving for x, you are to verify that x is in the numerator. If it is not, move it diagonally into the numerator on the other side of the equation. When the x term has only multiplication or division next to it, **then move only the x to the numerator.** For example:

$$4 = \frac{(9)(7) + 3}{(14)(x)(6)}$$

$$(x)(4) = \frac{(9)(7) + 3}{(14)(6)}$$

However, if addition or subtraction is connected to x, **the entire string of terms must be moved to the numerator along with x.** For example:

$$0.051 = \frac{975 + 940}{x + 19,690}$$

$$(x + 19,690)(0.051) = 975 + 940$$
$$(x + 19,690)(0.051) = 1915$$

In **Step 3**, terms must be moved, as necessary, so that x is by itself. (Since 19,690 is connected to x by an addition sign, it will temporarily be considered part of the x term.)

$$x + 19,690 = \frac{1915}{0.051}$$
$$x + 19,690 = 37,549$$

Now the addition term can be moved to the other side of the equation, leaving x by itself:

$$x = 37,549 - 19,690$$
$$x = \boxed{17,859}$$

PRACTICE PROBLEMS 2.4: Solving for x—Advanced Problems

1. $4x + 15 = 433$

$x =$ _____

6. $\dfrac{(6920)\,(8.34)\,(x) + (4610)\,(8.34)\,(0.07)}{57,800 + 38,120} = 23.4$

$x =$ _____

2. $\dfrac{(6 + x)\,(4)}{2} = 32$

$x =$ _____

7. $\dfrac{(2610)\,(0.44)\,(8.34)}{(5800)\,(x)\,(8.34) + (20)\,(1.6)\,(8.34)} = 8.2$

$x =$ _____

3. $0.051 = \dfrac{2150 + 792}{x + 28,540}$

$x =$ _____

8. $\dfrac{(6620)\,(8.34)\,(0.04) + (4480)\,(8.34)\,(0.065)}{58,100 + x} = 0.048$

$x =$ _____

4. $0.048 = \dfrac{x + 940}{19,410 + 20,490}$

$x =$ _____

9. $\dfrac{(x + 5.5)\,(3.5)\,(145)}{2} = 3426$

$x =$ _____

5. $\dfrac{(9 + 5)\,(x)}{2} = 35$

$x =$ _____

10. $\dfrac{(12,400)\,(0.95)\,(8.34)}{(6100)\,(0.032)\,(8.34) + (x)\,(1.95)\,(8.34)} = 9.1$

$x =$ _____

3 Fractions

Complete and score the following skills test. Each section should be scored separately in the box provided to the right. A score of 8 or above indicates you are sufficiently strong in that concept. A score of 7 or below indicates a review of that section is advisable.

3.1 Naming Fractions

Number Correct

❑ Write the denominator that corresponds to each of the following figures (write answer in circle).

1. =   2.  =   3.  =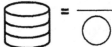

❑ Which figure best represents the fraction given?

4. $\frac{1}{2}$:   ANS____

      (a)   (b)   (c)   (d)

5. $\frac{1}{8}$: 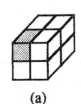   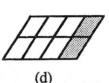   ANS____

      (a)   (b)   (c)   (d)

❑ In the following problems, write the fraction that represents the bold or shaded area.

6.  ANS____   7. 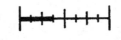 ANS____

8.  ANS____   9. ANS____

10.  ANS____

3.2 Equivalent Fractions

Number Correct

❑ Write two equivalent fractions that represent the shaded area in the figure below.

1.   ANS_____

_____

❑ Give an equivalent fraction (using multiplication) for each fraction listed below.

2. $\dfrac{3}{5}=$ **3.** $\dfrac{1}{7}=$ **4.** $\dfrac{9}{11}=$

❑ Give an equivalent fraction (using division) for each fraction listed below.

5. $\dfrac{10}{18}=$ **6.** $\dfrac{6}{36}=$ **7.** $\dfrac{16}{56}=$

❑ Are the pairs of fractions shown below equivalent fractions? If yes, what is the cross multiplication product?

8. $\dfrac{2}{3}=\dfrac{96}{144}$ ANS_____   **9.** $\dfrac{3}{14}=\dfrac{33}{152}$ ANS_____

Cross Product_____   Cross Product_____

10. $\dfrac{4}{7}=\dfrac{380}{665}$ ANS_____

Cross Product_____

3.3 Reducing Fractions

Number Correct

❑ Reduce each fraction to lowest terms.

1. $\dfrac{6}{8}=$ ____   **2.** $\dfrac{16}{20}=$ ____   **3.** $\dfrac{9}{12}=$ ____   **4.** $\dfrac{15}{25}=$ ____

5. $\dfrac{20}{24}=$ ____   **6.** $\dfrac{7}{19}=$ ____   **7.** $\dfrac{72}{81}=$ ____   **8.** $\dfrac{132}{352}=$ ____

9. $\dfrac{16}{52}=$ ____   **10.** $\dfrac{17}{30}=$ ____

3.4 Lowest Common Denominators

❑ Find the lowest common denominator for each group of fractions and rewrite the fractions using the LCD.

1. $\frac{2}{3}$, $\frac{4}{5}$ ANS_____

2. $\frac{5}{8}$, $\frac{7}{12}$ ANS_____

3. $\frac{1}{6}$, $\frac{3}{4}$ ANS_____

4. $\frac{1}{8}$, $\frac{16}{20}$ ANS_____

5. $\frac{2}{9}$, $\frac{1}{12}$ ANS_____

6. $\frac{1}{10}$, $\frac{43}{80}$ ANS_____

7. $\frac{1}{4}$, $\frac{3}{5}$, $\frac{1}{2}$ ANS_____

8. $\frac{2}{3}$, $\frac{3}{4}$, $\frac{1}{6}$ ANS_____

9. $\frac{7}{10}$, $\frac{1}{2}$, $\frac{3}{4}$ ANS_____

10. $\frac{2}{3}$, $\frac{7}{8}$, $\frac{5}{6}$ ANS_____

3.5 Improper Fractions and Mixed Numbers

❑ Write a mixed number for the part that is shaded. Reduce fractions to lowest terms.

1. 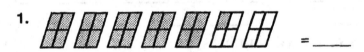 = _____

2.  = _____

❑ Write each mixed number as an improper fraction.

3. $6\frac{7}{8}$ = _____

4. $12\frac{2}{7}$ = _____

5. $5\frac{3}{5}$ = _____

6. $26\frac{2}{3}$ = _____

❑ Write each improper fraction as a whole number or mixed number in lowest terms.

7. $\frac{14}{5}$ = _____

8. $\frac{27}{8}$ = _____

9. $\frac{48}{7}$ = _____

10. $\frac{18}{10}$ = _____

3.6 Addition or Subtraction of Fractions or Mixed Numbers

Number Correct

❏ Add or subtract, as indicated. Reduce answers to lowest terms.

1. $\dfrac{5}{8} + \dfrac{3}{10} =$ _____

2. $5\dfrac{1}{8} - 2\dfrac{3}{4} =$ _____

3. $\dfrac{3}{4} + \dfrac{4}{7} =$ _____

4. $7\dfrac{4}{5} + 12\dfrac{2}{5} + 3\dfrac{1}{5} =$ _____

5. $3\dfrac{1}{4} - \dfrac{1}{9} =$ _____

6. $\dfrac{1}{5} - \dfrac{1}{8} =$ _____

7. $\dfrac{5}{6} + \dfrac{1}{12} =$ _____

8. $10\dfrac{1}{3} - 5\dfrac{3}{5} =$ _____

9. If a treatment plant receives 1/5 of the daily flow from District 1 and 2/7 of the daily flow from District 2, what fraction of the daily flow is contributed by these two districts?

ANS _____

10. Four sewer mains feed into a treatment plant. If three of the mains contribute relative flows of 1/3, 1/6, and 1/24, what is the fractional flow contributed by the fourth main?

ANS _____

3.7 Multiplication of Fractions or Mixed Numbers

Number Correct

❏ Multiply as indicated, using cancellation of common factors when possible. Reduce answers to lowest terms.

1. $\dfrac{3}{8} \times \dfrac{1}{7} =$ _____

2. $\dfrac{5}{6} \times \dfrac{7}{9} =$ _____

3. $\dfrac{9}{10} \times 1\dfrac{1}{4} =$ _____

4. $8\dfrac{1}{2} \times \dfrac{2}{3} =$ _____

5. $\dfrac{1}{5} \times 2 =$ _____

6. $\dfrac{13}{9} \times \dfrac{3}{26} =$ _____

7. $1\dfrac{1}{9} \times 7\dfrac{5}{10} =$ _____

8. $52 \times \dfrac{1}{3} =$ _____

Continued on next page...

3.7 Multiplication of Fractions or Mixed Numbers—Continued

9. Water fills a tank to 4/5 of its capacity. If the capacity of the tank is 35,000 cu ft, how many cu ft of water are in the tank?

ANS _____

10. On a particular day, the flow to the treatment plant was 7 million gallons. If 1/20 of the flow was industrial waste, how many million gallons of flow were industrial waste?

ANS _____

3.8 Division by Fractions or Mixed Numbers

Number Correct

❑ Divide the fractions and mixed numbers given below. Reduce answers to lowest terms.

1. $\dfrac{15}{16} \div \dfrac{5}{8} =$ _____

2. $\dfrac{3}{4} \div \dfrac{7}{9} =$ _____

3. $4 \div \dfrac{4}{12} =$ _____

4. $2\dfrac{1}{2} \div \dfrac{3}{2} =$ _____

5. $7\dfrac{2}{9} \div 5\dfrac{1}{3} =$ _____

6. $\dfrac{5}{6} \div \dfrac{2}{9} =$ _____

7. $\dfrac{11}{12} \div \dfrac{5}{14} =$ _____

8. $10 \div \dfrac{5}{6} =$ _____

9. $16\dfrac{2}{3} \div \dfrac{3}{4} =$ _____

10. $250 \div \dfrac{1}{4} =$ _____

3.9 Combined Calculations with Fractions

❑ Complete the problems shown below. Reduce answers to lowest terms. (Each problem is worth 2 points.)

1. $\dfrac{8\frac{3}{8} \times 5}{\frac{1}{4}} =$ _____

2. $\dfrac{\frac{3}{5} + \frac{6}{14}}{5 + \frac{1}{3}} =$ _____

3. $\dfrac{1\frac{1}{2} \times \frac{3}{7} \times \frac{1}{3}}{\frac{3}{8} \times 10} =$ _____

4. $\dfrac{\frac{1}{6} + \frac{2}{3} + \frac{4}{6}}{\frac{5}{12} - \frac{1}{3}} =$ _____

5. $\dfrac{1\frac{2}{3} \times \frac{3}{4}}{\frac{5}{9} + 6} =$ _____

3.1 NAMING FRACTIONS

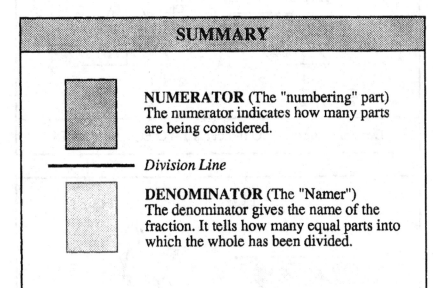

SUMMARY

NUMERATOR (The "numbering" part)
The numerator indicates how many parts are being considered.

Division Line

DENOMINATOR (The "Namer")
The denominator gives the name of the fraction. It tells how many equal parts into which the whole has been divided.

The concept of fractions was developed thousands of years ago so that portions of a whole object could be counted, recorded or perhaps shared equally.

The idea of **equal portions** is fundamental to the concept of fractions. In fact, it is how we name fractions.

THE DENOMINATOR

The denominator of a fraction gives the **name of the fraction** (halves, thirds, fourths, fifths, twentieths, etc.). It tells us how many equal parts into which the whole has been divided.

A denominator of two indicates that the whole has been divided into two equal parts.

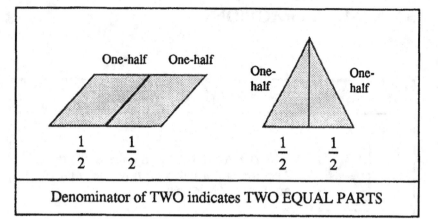

Denominator of TWO indicates TWO EQUAL PARTS

And if the denominator of the fraction is 3, the whole has been divided into three equal parts.

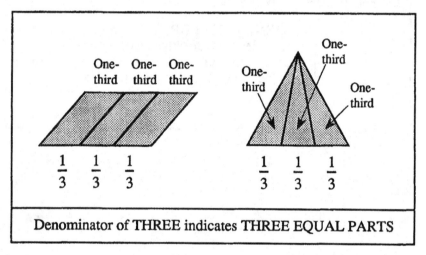

Denominator of THREE indicates THREE EQUAL PARTS

THE NUMERATOR

The top of the fraction, the numerator, indicates the **number of equal parts** that are of interest, as illustrated by the bold or shaded portions of the figures to the right.

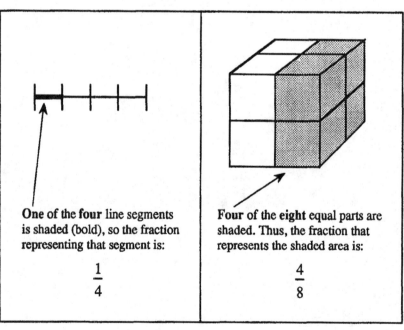

One of the **four** line segments is shaded (bold), so the fraction representing that segment is:

$$\frac{1}{4}$$

Four of the **eight** equal parts are shaded. Thus, the fraction that represents the shaded area is:

$$\frac{4}{8}$$

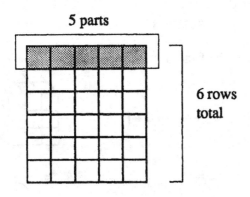

5 parts

6 rows total

There are **5 parts** in the row. Since there are a total of **6 rows** (each one having 5 parts), the total number of parts is:

$$\text{(5 }\underline{\text{parts}}\text{)(6 rows)} = 30 \text{ parts}$$
$$\text{row}$$

HOW MANY PARTS?

To determine how many parts a square or rectangle has been divided into, count how many "parts" are shown in the top row and then multiply by the total number of rows.

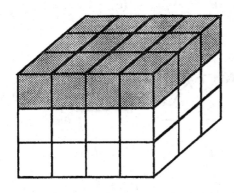

There are 4 parts in the row. Since there are a total of 3 rows in the top layer (each one having 4 parts), the number of parts in the <u>top layer</u> is: 3 x 4 = 12. However, in this example there are <u>3 layers</u>. Therefore, the total number of parts is:

$$\text{(12 }\underline{\text{parts}}\text{)(3 layers)} = 36 \text{ parts}$$
$$\text{layer}$$

A similar calculation may be used when a box-type object has been divided into parts.

PARTS OF A GROUP

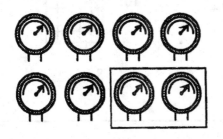

$$= \frac{2}{8}$$

Fractions may also be used to indicate parts of a group as well as portions of whole objects. In the example shown to the left, two of the eight meters are defective. The fraction that represents the number of defective meters is 2/8.

PRACTICE PROBLEMS 3.1: Naming Fractions

❏ Write the denominator that corresponds to each of the following figures:

1. = ──── 〇   **2.** = ──── 〇   **3.** = ──── 〇   **4.** = ──── 〇

❏ Which figure represents the fraction given?

5. $\frac{1}{4}$:   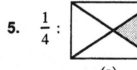   ANS _____

(a)   (b)   (c)   (d)

6. $\frac{6}{10}$:   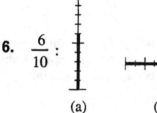 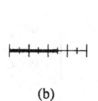 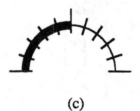   ANS _____

(a)   (b)   (c)   (d)

7. $\frac{2}{3}$:   ANS _____

(a)   (b)   (c)   (d)

❏ In the following problems, write the fraction that represents the shaded area.

8.   **9.**   **10.**

ANS _____   ANS _____   ANS _____

3.2 EQUIVALENT FRACTIONS

<table>
<tr><td colspan="3" align="center">**SUMMARY**</td></tr>
<tr><td colspan="3">

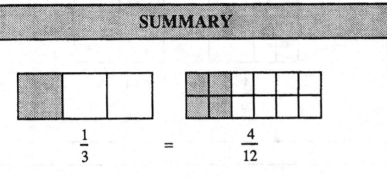

$$\frac{1}{3} \quad = \quad \frac{4}{12}$$

1. To obtain equivalent fractions, multiply or divide the numerator and denominator by the same number.

2. Cross multiply to determine if two fractions are equivalent fractions.

</td></tr>
</table>

Many times fractions with different numerators and denominators refer to the same portion of the whole. These fractions are called **equivalent fractions**.

EQUAL PARTS OF THE WHOLE

The shaded areas representing the fractions 4/16 and 1/4 are equal areas. In fact, they represent the same portion of the whole object. Fractions that represent equal parts of the whole are called equivalent fractions.

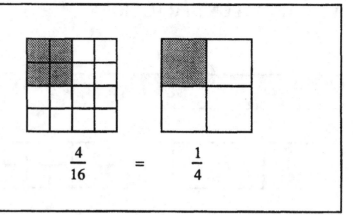

$$\frac{4}{16} \quad = \quad \frac{1}{4}$$

EQUIVALENT FRACTIONS BY MULTIPLYING

One way to obtain equivalent fractions is to **multiply the numerator and denominator by the same number**. This does not change the value of the original fraction since any number over itself is equal to one.

Each of these fractions 4/11, 12/33, 28/77, 36/99, and 48/132 are equivalent fractions since they represent the same portion of the whole.

Example 1: (Equivalent Fractions)
❏ Find an equivalent fraction of 4/11.

$$\frac{4}{11} \longrightarrow \frac{?}{?}$$

$$\frac{4}{11} \times \frac{3}{3} = \frac{12}{33}$$

Multiplying the numerator and denominator by 3 does not change the value of the fraction 4/11 since 3/3 is the same as 1. And 4/11 x 1 = 4/11.

4/11 and 12/33 are equivalent fractions. Other equivalent fractions are shown below.

$$\frac{4}{11} \times \frac{7}{7} = \frac{28}{77}$$

$$\frac{4}{11} \times \frac{9}{9} = \frac{36}{99}$$

$$\frac{4}{11} \times \frac{12}{12} = \frac{48}{132}$$

Example 2: (Equivalent Fractions)
❑ Find an equivalent fraction of 156/204.

$$\frac{156}{204} \longrightarrow \frac{?}{?}$$

$$\frac{156}{204} \div \frac{2}{2} = \frac{78}{102}$$

156/204 and 78/102 are equivalent fractions. Other equivalent fractions may be obtained by continuing to divide.

$$\frac{78}{102} \div \frac{2}{2} = \frac{39}{51}$$

$$\frac{39}{51} \div \frac{3}{3} = \frac{13}{17}$$

EQUIVALENT FRACTIONS BY DIVIDING

Another way to obtain equivalent fractions is to **divide the numerator and denominator by the same number.**

156/204, 78/102, 39/51, and 13/17 are equivalent fractions and represent the same portion of the whole.

Example 3: (Equivalent Fractions)
❑ Are 1/8 and 5/32 equivalent fractions?

$$1 \times 32 = 32 \qquad\qquad 8 \times 5 = 40$$

$$\frac{1}{8} = \frac{5}{32}$$

No, they are not equivalent fractions since the products of cross multiplication are not equal.

CROSS MULTIPLY TO VERIFY EQUIVALENT FRACTIONS

Given any two fractions, you can determine if they are equivalent fractions by simply cross multiplying, as shown in Example 3. **If the products of the cross multiplication are the same,** then the two fractions are indeed equivalent fractions.

PRACTICE PROBLEMS 3.2: Equivalent Fractions

❑ Write two equivalent fractions that represent the shaded areas in each figure below.

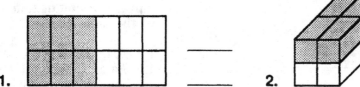

1. _____   **2.** _____

❑ Give two equivalent fractions (using multiplication) for the fraction 5/7.

3. _____   _____

❑ Find two equivalent fractions (using division) for the fraction 132/231.

4. _____   _____

❑ Are the pairs of fractions shown below equivalent fractions? If yes, what is their cross product?

5. $\dfrac{2}{8} = \dfrac{6}{24}$ _____

Cross Product _____

7. $\dfrac{5}{6} = \dfrac{125}{150}$ _____

Cross Product _____

6. $\dfrac{3}{4} = \dfrac{8}{12}$ _____

Cross Product _____

8. $\dfrac{3}{5} = \dfrac{27}{45}$ _____

Cross Product _____

3.3 REDUCING FRACTIONS

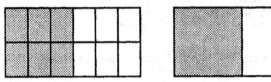

SUMMARY

$$\frac{6}{12} \begin{array}{c} \div \ 6 \\ \div \ 6 \end{array} \longrightarrow \frac{1}{2}$$

1. To reduce a fraction, divide the numerator and denominator by the **same** number.

2. Be sure to reduce to simplest terms. When reducing a fraction, you are trying to find the **smallest equivalent fraction** (see Sec. 3.2).

In making calculations with fractions, it is generally easier to work with fractions that have **smaller numerators and denominators**. For example, it is easier to work with the fraction 1/3 than with the equivalent fraction of 48/144. Although both represent the same portion of the whole, the larger fraction is much more cumbersome to use in calculations, and it is also more difficult to visualize its relative size. In order to make calculations easier, therefore, fractions are often **reduced to simplest terms** prior to making calculations with them.

DIVIDE NUMERATOR AND DENOMINATOR BY THE SAME NUMBER

To reduce a fraction to simpler terms, **divide the numerator and denominator by the same number.**

Continue to reduce the fraction until the numerator and denominator can no longer be divided evenly by any other number (except 1). When you can no longer divide, it has been reduced to lowest or simplest terms.

DIVIDE BY THE LARGEST NUMBER FIRST

It is desirable to divide the numerator and denominator by the **largest number that comes to mind**. This will help reduce the fraction in the fewest possible steps. For example, if you can see that the numerator and denominator of a fraction are both divisible by 10, it is quickest to divide them by 10 rather than divide them by 5 and then again by 2. The final answer will be the same but there are **fewer division steps when larger numbers are used.**

Example 1: (Reducing Fractions)
❏ Reduce the fraction 8/30.

Both 8 and 30 are even numbers. They are both divisible by 2:

$$\frac{8}{30} \div \frac{2}{2} = \frac{4}{15}$$

8/30 and 4/15 are equivalent fractions and therefore represent the same part of the whole. However, 4/15 is easier to use in calculations than 8/30 and is preferred.

Example 2: (Reducing Fractions)
❏ Reduce 45/75 to simplest terms.

$$\frac{45}{75} \div \frac{5}{5} = \frac{9}{15} \div \frac{3}{3} = \frac{3}{5}$$

Example 3: (Reducing Fractions)
❏ Reduce the fraction 486/1491 to simplest terms.

Consider the **largest numbers first**: 10, 5, 3, and 2. Ten will not divide evenly into either 486 or 1491, nor will 5, according to the divisibility rules shown on the next page. However, three will divide evenly into both numbers:

$$\frac{486}{1491} \div \frac{3}{3} = \frac{162}{497}$$

RULES OF DIVISIBILITY	
Division by 2	When a number is evenly divisible by 2, the **last digit in the number will be an even number.** 2796 —The last digit (6) is EVEN. Therefore, 2796 is divisible by 2. 12,601 —The last digit (1) is ODD. Therefore, 12,601 is not divisible by 2.
Division by 3	When a number is evenly divisible by 3, the **sum of the digits will also be divisible by 3.** 123 — The sum of the digits (1+2+3) is equal to 6. Since 6 is divisible by 3 the number 123 will also be divisible by 3. 509 — The sum of the digits (5+0+9) is equal to 14. Since 14 is not divisible by 3, the number 509 will also not be divisible by 3.
Division by 5	When a number is evenly divisible by 5, the **last digit will be either 5 or 0.** 740 — The last digit (0) fulfills the requirement of divisibility, and thus 740 is divisible by 5. 223 — The last digit (3) is not a 5 or 0. Thus the number 223 is not divisible by 5.

In reducing fractions with large numerators and denominators, it is useful to know whether the number is divisible by 2, 3, or 5. These are the numbers used most often in reducing fractions to lowest terms.

PRACTICE PROBLEMS 3.3: Reducing Fractions

❑ Reduce each fraction to lowest terms:

1. $\dfrac{18}{24}$ = _____

2. $\dfrac{20}{44}$ = _____

3. $\dfrac{6}{15}$ = _____

4. $\dfrac{25}{75}$ = _____

5. $\dfrac{21}{35}$ = _____

6. $\dfrac{18}{56}$ = _____

7. $\dfrac{12}{40}$ = _____

8. $\dfrac{6}{16}$ = _____

9. $\dfrac{90}{120}$ = _____

10. $\dfrac{72}{189}$ = _____

3.4 FINDING THE LOWEST COMMON DENOMINATOR

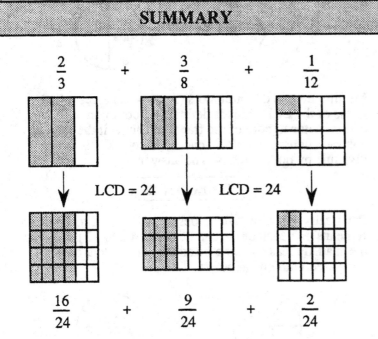

<table>
<tr><td colspan="5" align="center">**SUMMARY**</td></tr>
</table>

$$\frac{2}{3} \quad + \quad \frac{3}{8} \quad + \quad \frac{1}{12}$$

LCD = 24   LCD = 24

$$\frac{16}{24} \quad + \quad \frac{9}{24} \quad + \quad \frac{2}{24}$$

1. The lowest common denominator for several fractions is the **smallest number evenly divisible** by all denominators.

2. The simplest method to finding the LCD is often effective when the denominators are small:

   • Ask "What is the smallest number (besides 1) that all the denominators can divide into evenly?" (*Hint*: sometimes that number can be found by multiplying the denominators together.)

3. To find the LCD using the **factoring method**:

   • Factor each denominator to lowest terms,
   • List the factors and determine the greatest number of times each factor occurs,
   • Multiply those factors together to obtain the LCD.

4. Once you have found the LCD, convert each fraction to a fraction with the LCD as the denominator. (Refer to Section 3.2—Equivalent Fractions.)

When attempting to add or subtract fractions with **unlike denominators**, the difficulty becomes apparent—the **"parts" to be added are different sizes.**

In the example shown to the left, there are 2 parts, 3 parts, and 1 part to be added, as shown by the shaded areas. Thus, there are a total of 6 parts. But what name can be given to the parts? Are they 6 thirds, or 6 eighths, or 6 twelfths? The answer is—they are none of these.

Each of these fractions must be converted to fractions with the same denominators ("common denominators"). When they have the same denominators, the "parts" are the same size and may therefore be added, subtracted, or compared with other similar parts.

Although fractions can have more than one common denominator, it is a good policy to use the lowest common denominator (LCD) when making calculations. The following section describes two methods for finding the LCD:

• The multiplication method, and

• The factoring method.

LCD—THE MULTIPLICATION METHOD

Of the two methods for finding the lowest common denominator (LCD), the multiplication method is perhaps the simplest. Although this method does not work well with large denominators or for problems with several small denominators, it should not be overlooked since **it can sometimes be the most direct method of finding the LCD.**

The multiplication method involves multiplying the denominators together. The product will always be a common denominator, however, it may not be the lowest common denominator. Therefore, after obtaining what we will call the **"first guess" LCD**, check to be sure it is the lowest common denominator. In effect, you are asking, "Is there a smaller number that all the denominators can divide evenly into as well?" (*Hint*: often that smaller number is one-half or one-third of the "first guess" LCD.)

USE EQUIVALENT FRACTIONS TO CONVERT TO FRACTIONS

The LCD represents the **new denominator**. All the fractions in the given problem are then converted to fractions with the same denominator—the LCD. This is done using the concept of equivalent fractions*. Once the fractions all have the same denominators, they can be added, subtracted, or compared as desired.

Example 1: (LCD—The Multiplication Method)
❏ Find the LCD for 1/3 and 1/4.

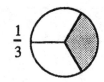 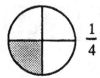

Multiplying the denominators (3 x 4 = 12) gives a first guess of the LCD. This is definitely a common denominator to both of the fractions, but is it the lowest common denominator? To check; 1/2 of 12 is 6. Is 6 divisible by both 3 and 4? The answer is no.

$$\boxed{\text{The LCD is 12.}}$$

Practically speaking, what does an LCD equal to 12 mean? It means that each circle can be divided into 12 parts, and the shaded areas can then be added, subtracted, or compared with one another:

$$\boxed{\dfrac{1}{3}} \begin{array}{c} \times\ 4 \\ \times\ 4 \end{array} = \boxed{\dfrac{4}{12}} \qquad \boxed{\dfrac{1}{4}} \begin{array}{c} \times\ 3 \\ \times\ 3 \end{array} = \boxed{\dfrac{3}{12}}$$

Example 2: (LCD—The Multiplication Method)
❏ Find the LCD for the following fractions: 5/8 and 4/5, then rewrite the fractions using the LCD.

Multiplying the denominators (5 x 8 = 40) gives us the "first guess" LCD. Is there another lower number that both denominators can divide into evenly? It does not appear so, therefore the LCD is 40.

$$\boxed{\text{LCD} = 40}$$

Now rewrite each fraction using the LCD:

$$\dfrac{5}{8} \begin{array}{c} \times\ 5 \\ \times\ 5 \end{array} = \boxed{\dfrac{25}{40}} \qquad \dfrac{4}{5} \begin{array}{c} \times\ 8 \\ \times\ 8 \end{array} = \boxed{\dfrac{32}{40}}$$

---

*Refer to Section 3.2—Equivalent Fractions.

Example 3: (LCD—The Multiplication Method)
❑ Find the LCD for 1/2, 2/5, and 2/3 then rewrite each fraction using the LCD.

To find the "first guess" LCD, multiply all three denominators together (2 x 5 x 3 = 30). So 30 is the "first guess" LCD. Is there another <u>smaller number</u> that is also evenly divisible by all three numbers? Perhaps 15 or 10? No, all three numbers do not evenly divide into either of these two numbers.

$$\boxed{\text{LCD} = 30}$$

Rewriting each fraction using the LCD:

$$\frac{1 \times 15}{2 \times 15} = \boxed{\frac{15}{30}} \qquad \frac{2 \times 6}{5 \times 6} = \boxed{\frac{12}{30}} \qquad \frac{2 \times 10}{3 \times 10} = \boxed{\frac{20}{30}}$$

Example 4: (LCD—The Multiplication Method)
❑ Find the LCD for 2/3, 5/6 and 1/4, then rewrite the fractions using the LCD.

The "first guess" LCD would be (3 x 6 x 4 = 72). However, it would appear that there is a lower number than 72 that all three denominators could divide into evenly. One half of 72 = 36. All three denominators divide evenly into 36. What about a lower number yet? One-third of 36 = 12, which is also evenly divisible by all three denominators. The LCD must therefore be 12.

$$\boxed{\text{LCD} = 12}$$

Rewriting the fractions using the LCD:

$$\frac{2 \times 4}{3 \times 4} = \boxed{\frac{8}{12}} \qquad \frac{5 \times 2}{6 \times 2} = \boxed{\frac{10}{12}} \qquad \frac{1 \times 3}{4 \times 3} = \boxed{\frac{3}{12}}$$

This example illustrates the **limitation of the multiplication method** in finding the LCD. It required multiplying the three denominators, then dividing the resulting number two times. The multiplication method is a more effective tool when you become familiar with the factoring method of finding the LCD. With experience, you will begin to anticipate which method works best for the problem at hand.

LCD—THE FACTORING METHOD

To find the lowest common denominator using the factoring approach:

1. **Factor each denominator** to lowest terms (use a "factor tree"),

2. **List the factors** represented in any factor tree and determine the greatest number of times each factor occurs, and

3. **Multiply those factors** together to obtain the LCD.

When factoring, you may select different numbers to begin the factoring, yet the final result will be the same. For example, if you had begun factoring 12 with the factors 3 and 4 (rather than 6 and 2 as shown in Example 5), the final factors will be the same.

Example 5: (LCD—The Factoring Method)
❑ Find the LCD for 11/12 and 7/16.

Step 1—Factor each denominator, using the factor tree.

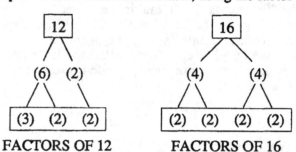

FACTORS OF 12   FACTORS OF 16

Step 2—After completing the factor trees, list all the factors shown in any factor tree. (Look at the last level of the factor tree only.)

Factor Shown	Greatest Number of Times That Factor Occurs in Any Tree
2 ⟶	2 occurs the <u>greatest</u> number of times in the 16 factor tree: (2)(2)(2)(2)
3 ⟶	3 occurs the <u>greatest</u> number of times in the 12 factor tree: (3)

Step 3—Multiply the factors in the "Greatest Times" list to obtain the LCD:

$$(2)(2)(2)(2)(3) = \boxed{48}$$
$$\text{LCD}$$

Example 6: (LCD—The Factoring Method)
❑ Find the LCD of 1/6, 5/24, and 3/8.

Step 1—Factor the denominators to lowest terms.

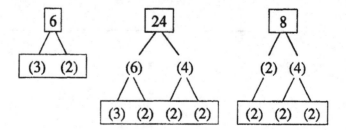

(Continued on next page)

Example 6: (Continued)

Step 2: List factors shown and the greatest number of times that factor is shown for any factor tree.

Factor Shown	Greatest Number of Times That Factor Occurs in Any Tree
2 ⟶	(2)(2)(2)
3 ⟶	(3)

Step 3—Multiply factors from the "Greatest Times" list.

$$(2)(2)(2)(3) = \boxed{24}$$
LCD

Example 7: (LCD—The Factoring Method)
❏ Find the LCD of 7/10, 13/15 and 5/12, then rewrite the fractions using the LCD.

Step 1—Factor the denominators.

  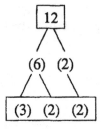

Steps 2 & 3—List the greatest number of times each factor occurs in any factor tree, and multiply the factors:

$$LCD = (2)(2)(3)(5)$$
$$= 60$$

Then rewrite fractions:

$$\frac{7 \times 6}{10 \times 6} = \boxed{\frac{42}{60}} \qquad \frac{13 \times 4}{15 \times 4} = \boxed{\frac{52}{60}} \qquad \frac{5 \times 5}{12 \times 5} = \boxed{\frac{25}{60}}$$

PRACTICE PROBLEMS 3.4: Lowest Common Denominators

❑ Use the factoring method to determine the lowest common
denominator for each group of fractions given below.

1. $\dfrac{1}{6}, \dfrac{2}{9}, \dfrac{5}{18}$ LCD = _____

2. $\dfrac{3}{5}, \dfrac{7}{8}, \dfrac{9}{16}$ LCD = _____

3. $\dfrac{3}{8}, \dfrac{7}{12}, \dfrac{5}{6}$ LCD = _____

4. $\dfrac{2}{3}, \dfrac{1}{4}, \dfrac{4}{5}$ LCD = _____

5. $\dfrac{3}{4}, \dfrac{7}{10}, \dfrac{5}{12}$ LCD = _____

❑ Find the LCD, then convert the fractions to like fractions.

6. $\dfrac{1}{2}, \dfrac{3}{8}, \dfrac{7}{12}$ = _____

7. $\dfrac{2}{3}, \dfrac{4}{9}, \dfrac{4}{12}$ = _____

8. $\dfrac{2}{5}, \dfrac{5}{6}, \dfrac{9}{15}$ = _____

9. $\dfrac{5}{12}, \dfrac{3}{16}, \dfrac{7}{8}$ = _____

10. $\dfrac{2}{3}, \dfrac{1}{2}, \dfrac{5}{6}$ = _____

3.5 IMPROPER FRACTIONS AND MIXED NUMBERS

<table>
<tr><td>

SUMMARY

1. To convert a **mixed number to an improper fraction**:

- **Multiply** the denominator and the whole number.

- **Add** the product obtained above to the numerator. This total will be the numerator of the improper fraction; the denominator will remain the same.

- **Reduce** the fraction to lowest terms.

$$4 \overset{\nearrow}{\underset{\nwarrow}{}} \frac{1}{5}$$

$$4\frac{1}{5} = \boxed{\frac{21}{5}}$$

2. To convert an **improper fraction to a mixed number**:

- **Reduce** the improper fraction, if possible.

- **Divide** the numerator by the denominator to obtain the whole number.

- Any **remainder becomes the numerator** of the fractional part of the mixed number. Use the same denominator used for the improper fraction.

- **Reduce** the fraction part of the mixed number, if possible.

</td></tr>
</table>

In Sections 2.1-2.4 we have worked exclusively with fractions whose numerators were less than the denominators. These fractions are called proper fractions. Examples of **proper fractions** are given below. Each of these fractions <u>represents a number less than one</u>:

$$\frac{1}{4}, \ \frac{2}{3}, \ \frac{6}{7}, \ \frac{43}{48}$$

An **improper fraction** has a numerator equal to or greater than the denominator. As shown in the examples given below, improper fractions represent numbers equal to or greater than one:

$$\frac{10}{3}, \ \frac{9}{8}, \ \frac{7}{7}, \ \frac{25}{2}$$

Mixed numbers, as the name suggests, are numbers comprised of both whole numbers and fractions. Examples of mixed numbers are given below:

$$2\frac{1}{2}, \ 5\frac{7}{8}, \ 15\frac{4}{7}, \ 3\frac{5}{6}$$

MIXED NUMBERS TO IMPROPER FRACTIONS

Often, in calculations dealing with addition, subtraction, multiplication, or division of fractions, a mixed number must be converted to an improper fraction before further calculation is possible.

The diagram below illustrates the mixed number 2-3/4.

$2\frac{3}{4} =$

Since the fractional part of the mixed number is <u>fourths</u>, the wholes can also be expressed as <u>fourths</u>:

$2\frac{3}{4} =$ $= \frac{11}{4}$

The mixed number 2-3/4 has been converted to <u>fourths</u>.

The mixed number 4-2/5 is shown in the diagram below.

$4\frac{2}{5} =$

Since the fractional part of the mixed number is <u>fifths</u>, the wholes can also be shown as <u>fifths</u>:

$4\frac{2}{5} =$ $= \frac{22}{5}$

Although diagrams may be useful in understanding the relationship between mixed numbers and improper fractions, it is not practical to make a diagram each time a mixed number is to be converted to an improper fraction. The mathematical conversion will use both multiplication and addition.

1. **Multiply** the denominator and the whole number.

2. **Add** the product obtained in Step 1 to the numerator. The total will be the numerator of the improper fraction. Use the same denominator.

3. **Reduce** the fraction, if possible.

Example 1: (Mixed Numbers to Fractions)
❑ Convert 2-3/4 to a fraction.

$$2 \underset{\nwarrow}{\overset{\nearrow}{}} \frac{3}{4}$$

Step 1—Multiplication
(4)(2) = 8

Step 2—Addition
Add 3 to the answer in Step 1.
8 + 3 = 11

The 11 indicates the total number of parts. Using the same denominator, the improper fraction is:

$$\boxed{\frac{11}{4}}$$

Example 2: (Mixed Numbers to Fractions)
❏ Express 6-6/8 as an improper fraction.

$$6 \overset{\nearrow}{\underset{\nwarrow}{}} \frac{6}{8}$$

Step 1—Multiplication
$(8)(6) = 48$

Step 2—Addition
Add 6 to the answer in Step 1.
$48 + 6 = 54$

The improper fraction is therefore:

$$\frac{54}{8}$$

Step 3—Reduce the fraction

$$\frac{54}{8} \begin{array}{c} \div \\ \div \end{array} \begin{array}{c} 2 \\ 2 \end{array} = \boxed{\frac{27}{4}}$$

MAKE IT EASY—REDUCE FIRST

Occasionally, the fraction portion of a mixed number can be reduced. **Reduce the fraction first,** before converting the mixed number to an improper fraction. This results in an easier conversion calculation.

In Example 2, 6/8 could have been reduced to 3/4. The conversion to an improper fraction would have been:

$$6\frac{3}{4} = \frac{27}{4}$$

The multiplication and addition steps involve smaller numbers and there is no reducing required at the end of the problem.

Example 3: (Mixed Numbers to Fractions)
❏ Convert 10-5/16 to an improper fraction.

$$10 \overset{\nearrow}{\underset{\nwarrow}{}} \frac{5}{16}$$

Step 1—Multiplication
$(6)(10) = 160$

Step 1—Addition
Add 5 to the answer in Step 1.
$160 + 5 = 165$

The improper fraction is:

$$\boxed{\frac{165}{16}}$$

IMPROPER FRACTIONS TO MIXED NUMBERS

Mixed numbers are most often converted to improper fractions when the numbers are to be used in multiplication or division problems. Sometimes, however, you will want to convert an improper fraction to a mixed number.

After completing multiplication or division by fractions, for example, it is often advisable to convert your answer to a mixed number. The size of a number is more easily understood when it is written as a mixed number (e.g., 10-19/23) than when that same number is written as an improper fraction (e,g., 249/23).

To convert from an improper fraction to a mixed number:

1. **Reduce** the improper fraction, if possible.

2. **Divide** the numerator by the denominator to obtain the whole number.

3. Any **remainder becomes the numerator** of the mixed number. Use the same denominator as given for the improper fraction.

4. **Reduce** the fraction part of the mixed number, if possible.

Example 4: (Fractions to Mixed Numbers)
❑ Convert 92/5 to a mixed number.

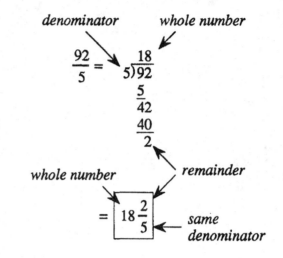

Example 5: (Fractions to Mixed Numbers)
❑ Express 125/15 as a mixed number.

First reduce the improper fraction:

$$\frac{125 \div 5}{15 \div 5} = \frac{25}{3}$$

Then divide:

$$\frac{25}{3} \quad = \quad 3\overline{)25} \begin{array}{l} 8 \leftarrow \textit{whole number} \\ \underline{24} \\ 1 \leftarrow \textit{remainder} \end{array}$$

$$= \boxed{8\,\frac{1}{3}}$$

Example 6: (Fractions to Mixed Numbers)
❏ Write 128/11 as a mixed number.

$$\frac{128}{11} \quad = \quad 11\overline{)128} \atop $$

$$\begin{array}{r} 11 \\ 11\overline{)128} \\ \underline{11} \\ 18 \\ \underline{11} \\ 7 \end{array}$$

$$= \boxed{11\frac{7}{11}}$$

Example 7: (Fractions to Mixed Numbers)
❏ Convert 275/7 to a mixed number.

$$\frac{275}{7} \quad = \quad \begin{array}{r} 39 \\ 7\overline{)275} \\ \underline{21} \\ 65 \\ \underline{63} \\ 2 \end{array}$$

$$= \boxed{39\frac{2}{7}}$$

PRACTICE PROBLEMS 3.5: Improper Fractions and Mixed Numbers

❑ Write a mixed number for the part that is shaded.
Reduce the fractional part to lowest terms.

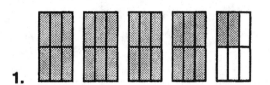   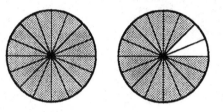

1. _____   **2.** _____

❑ Write each mixed number as an improper
fraction.

3. $2\dfrac{5}{8}$ = _____   **4.** $4\dfrac{4}{5}$ = _____   **5.** $3\dfrac{1}{9}$ = _____   **6.** $16\dfrac{3}{4}$ = _____

❑ Write each improper fraction as a whole number
or mixed number in lowest terms.

7. $\dfrac{25}{14}$ = _____   **8.** $\dfrac{98}{11}$ = _____   **9.** $\dfrac{19}{8}$ = _____   **10.** $\dfrac{19}{6}$ = _____

3.6 ADDITION OR SUBTRACTION OF FRACTIONS OR MIXED NUMBERS

Fractions can be added and subtracted directly only when they have the **same denominator** (like fractions). If they have **different denominators** (unlike fractions), they must first be converted to like fractions using the lowest common denominator (LCD).

Since mixed numbers include fractions as part of the number, they are discussed in this section as well.

SUMMARY

Like Fractions

$$\frac{1}{6} + \frac{4}{6} = \frac{5}{6}$$

When adding or subtracting like fractions (fractions with the same denominator):

- **Add or subtract the numerators,** then use the same denominator.

- **Reduce** your answer to lowest terms.

Unlike Fractions

$$\frac{7}{12} - \frac{1}{3} = ?$$

$$\frac{7}{12} - \frac{4}{12} = \frac{3}{12}$$

$$\text{or} = \frac{1}{4}$$

When adding or subtracting unlike fractions (fractions with different denominators):

- **Find the LCD*** and convert unlike fractions to like fractions.

- **Continue** as for like fractions, described above.

Mixed Numbers

$$3\frac{1}{4} + 2\frac{2}{4} = 5\frac{3}{4}$$

- **Restate the mixed number** if needed. (Sometimes borrowing is required for subtracting problems.)

- **Add or subtract fractions** according to the steps listed for like and unlike fractions.

- **Add or subtract whole numbers.**

- **Reduce** fractions to lowest terms.

---

*Refer to Section 3.4 for a discussion of lowest common denominators.

ADDING OR SUBTRACTING LIKE FRACTIONS

When adding or subtracting fractions with the **same denominator,** you should:

1. **Add or subtract** the numerators, then use the same denominator.

2. **Reduce** the answer to lowest terms.

There is no calculation involving the denominator since the denominator merely indicates the size of the objects being counted, as illustrated in the example to the right.

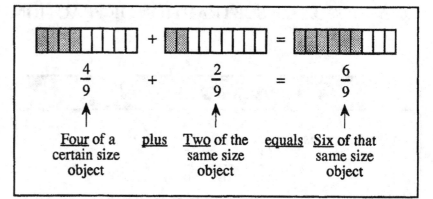

$$\frac{4}{9} \quad + \quad \frac{2}{9} \quad = \quad \frac{6}{9}$$

Four of a certain size object   plus   Two of the same size object   equals   Six of that same size object

Example 1: (Addition or Subtraction of Fractions)
❑ Add the following fractions: 1/7 and 4/7.

Step 1—Add numerators.

$$\frac{1}{7} + \frac{4}{7} = \boxed{\frac{5}{7}}$$

Step 2—Reduce the fraction.
This fraction is already in lowest terms

$$\boxed{\frac{5}{7}}$$

Example 2: (Addition or Subtraction of Fractions)
❑ Add the following fractions: 17/32 and 3/32.

Step 1—Add numerators.

$$\frac{17}{32} + \frac{3}{32} = \frac{20}{32}$$

Step 2—Reduce the fraction.

$$\frac{20 \div 4}{32 \div 4} = \boxed{\frac{5}{8}}$$

DIFFERENT DENOMINATORS MEAN DIFFERENT SIZED PARTS

1	$\frac{1}{2}$	$\frac{1}{3}$	$\frac{1}{4}$	$\frac{1}{8}$
The whole.	The whole is divided into 2 parts.	The whole is divided into 3 parts.	The whole is divided into 4 parts.	The whole is divided into 8 parts.

ADDING OR SUBTRACTING UNLIKE FRACTIONS

In many cases, fractions to be added or subtracted have **different denominators**. When this occurs, you <u>cannot</u> just add or subtract numerators. Why not?

To answer this, let's review the concept of fractions. The denominator of a fraction indicates the number of equal parts into which the whole has been divided. In other words, the denominator tells us about the "size of the pieces" into which the whole was divided.

Different sized objects cannot be added directly, because this would be similar to adding "one apple plus two oranges". To resolve this problem, use the lowest common denominator to <u>restate</u> the fractions as like fractions.

DIFFERENT SIZES CANNOT BE ADDED

$$\frac{1}{3} + \frac{1}{2} = \frac{2}{?}$$

The LCD divides the objects into the <u>same number of equal parts</u> so that they can then be added or subtracted. In this illustration, the LCD = 6.

$$\frac{2}{6} + \frac{3}{6} = \frac{5}{6}$$

Therefore, to add or subtract unlike fractions:

1. **Find the LCD** and convert unlike fractions to like fractions.

2. **Add or subtract** the numerators, then use the same denominator.

3. **Reduce** your answer.

Example 3: (Adding or Subtracting Fractions)
❑ Subtract 2/9 from 5/7.

First, find the LCD and restate the fractions.

$$LCD = (7)(3)(3)$$
$$= 63$$

$$\frac{5 \times 9}{7 \times 9} = \boxed{\frac{45}{63}} \qquad \frac{2 \times 7}{9 \times 7} = \boxed{\frac{14}{63}}$$

The problem can now be restated and completed:

$$\frac{45}{63} - \frac{14}{63} = \boxed{\frac{31}{63}}$$

ADDING OR SUBTRACTING MIXED NUMBERS

When adding or subtracting mixed numbers:

1. **Restate** the mixed number, if needed for subtraction. (Borrowing from the whole number requires this restating of the mixed number.)

2. **Add or subtract fractions** according to the steps described for like and unlike fractions.

3. **Add or subtract the whole numbers.**

4. **Reduce** any fractions to lowest terms.

Example 4: (Adding or Subtracting Mixed Numbers)
❏ Add the following mixed numbers: 2-3/8 and 1-1/8.

$$2\frac{3}{8}$$
$$+\ 1\frac{1}{8}$$
$$\overline{\ 3\frac{4}{8}}$$

— First add fractions
— Then add whole numbers

Reduce the fractional part when possible:

$$3\frac{4}{8}\ =\ \boxed{3\frac{1}{2}}$$

Example 5: (Adding or Subtracting Mixed Numbers)
❏ Complete the following problem:

$$7\frac{1}{2}$$
$$-\ 5\frac{3}{8}$$

Before subtracting the fractions, an LCD should be used to convert these unlike fractions to like fractions:

$$\text{LCD} = 8$$

$$\frac{1}{2} \times \frac{4}{4} = \boxed{\frac{4}{8}}\quad \boxed{\frac{3}{8}}$$

The problem can now be rewritten:

$$7\frac{4}{8}$$
$$-\ 5\frac{3}{8}$$
$$\overline{\ 2\frac{1}{8}}$$

— First subtract fractions
— Then subtract whole numbers

Example 6: (Adding or Subtracting Mixed Numbers)
❑ Complete the following subtraction: 9-2/7 minus 5-5/7.

$$9\frac{2}{7}$$

$$-\;5\frac{5}{7}$$

Restate mixed number.
Normally on a problem such as this, you would first subtract fractions. However, in this example you have a smaller fraction on the top. (You cannot subtract 5/7 when you have only 2/7 to begin with.) Therefore, you will need to borrow a whole number and convert it into a fraction:

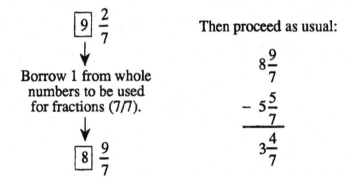

Borrow 1 from whole numbers to be used for fractions (7/7).

Then proceed as usual:

$$8\frac{9}{7}$$

$$-\;5\frac{5}{7}$$

$$3\frac{4}{7}$$

BORROWING FROM WHOLE NUMBERS

When the fraction on the top is smaller than the fraction on the bottom, you will have to **borrow from the whole number to increase the size of the top fraction,** as demonstrated in Example 6.

Example 7: (Adding or Subtracting Mixed Numbers)
❑ Subtract as indicated: 15-1/5 minus 3-5/6

First, convert unlike fractions to like fractions, using the LCD: (LCD = 30)

$$\frac{1}{5}\times\frac{6}{6}=\boxed{\frac{6}{30}}\qquad\frac{5}{6}\times\frac{5}{5}=\boxed{\frac{25}{30}}$$

$$15\frac{6}{30}\xrightarrow[\text{fraction}]{\substack{\text{borrow to}\\\text{increase}}}14\frac{36}{30}$$

$$-\;3\frac{25}{30}\qquad\qquad-\;3\frac{25}{30}$$

$$11\frac{11}{30}$$

PRACTICE PROBLEMS 3.6: Addition and Subtraction of Fractions or Mixed Numbers

❏ Add or subtract as indicated. Reduce your answers to lowest terms.

1. $\dfrac{8}{9} + \dfrac{7}{8} =$ _____

2. $\dfrac{1}{2} + \dfrac{3}{4} + \dfrac{5}{6} =$ _____

3. $6\dfrac{1}{3} + 3\dfrac{1}{4} =$ _____

4. $\dfrac{1}{2} + \dfrac{2}{3} =$ _____

5. $\dfrac{1}{4} + \dfrac{5}{6} =$ _____

6. $11\dfrac{1}{7} - 4\dfrac{1}{8} =$ _____

7. $\dfrac{1}{6} + \dfrac{2}{3} - \dfrac{1}{2} =$ _____

8. $\dfrac{9}{7} - \dfrac{24}{78} =$ _____

9. $18\dfrac{8}{13} - 17\dfrac{1}{2} =$ _____

10. $20\dfrac{1}{3} + 15\dfrac{4}{21} =$ _____

3.7 Multiplication of Fractions or Mixed Numbers

SUMMARY

Two simple fractions may be used to illustrate multiplication of fractions. To understand the concept, the **multiplication sign is replaced by the word "of"**:

$$\frac{1}{3} \times \frac{1}{2} =$$

$$\frac{1}{3} \text{ of } \frac{1}{2} =$$

Let's look at this statement using an illustration:

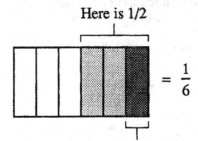

Here is 1/2

$$= \frac{1}{6}$$

Here is 1/3 of that 1/2

When **multiplying fractions**:

- **Multiply** the numerators

- **Multiply** the denominators

- **Reduce** fractions to lowest terms

$$\frac{1}{3} \times \frac{1}{2} = \boxed{\frac{1}{6}}$$

When multiplying **mixed numbers**:

- **Convert** the mixed numbers to improper fractions

- **Multiply** the numerators and denominators

- **Reduce** fractions to lowest terms

In addition or subtraction of fractions, we are adding or subtracting equally sized parts to obtain a total. But what are we really doing when one fraction is multiplied by another?

The key to understanding what occurs in these problems is found by replacing the multiplication sign by the word "of," as illustrated in the diagram to the left.

MULTIPLYING FRACTIONS

When multiplying fractions:

1. **Multiply** the numerators,

2. **Multiply** the denominators,

3. **Reduce** fractions to lowest terms.

Example 1: (Multiplication of Fractions)

❑ Complete the following multiplication: 7/10 x 5/8.

$$\text{Multiply Numerators} \longrightarrow \atop \text{Multiply Denominators} \longrightarrow \quad \frac{7}{10} \times \frac{5}{8} = \frac{35}{80}$$

Reduce fraction to lowest terms:

$$\frac{35 \div 5}{80 \div 5} = \boxed{\frac{7}{16}}$$

SHORTCUT WHEN MULTIPLYING FRACTIONS

When multiplying fractions, you can simplify the process considerably by dividing out or **cancelling common factors*** in the numerators and denominators **before multiplying them.** Example 2 illustrates multiplication of fractions with and without cancellation. With more complex problems, dividing out factors becomes even more important.

When dividing out factors, you must divide out one factor in the numerator for every factor divided out in the denominator. This keeps the fractions <u>in balance</u>.

You may have noticed in Example 2 that the cancellation or dividing out process is much like reducing a fraction to the lowest terms**. When you divide out a number in <u>both</u> the numerator and denominator, you have not changed the <u>value</u> of that fraction.

Cancellation is **to be used only in multiplication and division of fractions,** <u>never</u> in addition and subtraction of fractions.

Example 2: (Multiplication of Fractions)

❑ Multiply the following fractions: 3/5 x 45/7.

In this problem multiplication of fractions with and without cancellation will be illustrated.

Multiplication without cancellation:

$$\frac{3}{5} \times \frac{45}{7} = \frac{135}{35}$$

Reduce to lowest terms:

$$\frac{135 \div 5}{35 \div 5} = \frac{27}{7}$$

$$= \boxed{3\frac{6}{7}}$$

Multiplication with cancellation:

$$\frac{3}{\underset{1}{\cancel{5}}} \times \frac{\overset{9}{\cancel{45}}}{7}$$

First check for common factors. You are looking for a number which will divide evenly into one of the numerators <u>and</u> one of the denominators. Notice that <u>both 5 and 45</u> are evenly divisible by 5.

So the problem is now:

$$\frac{3}{1} \times \frac{9}{7} = \frac{27}{7}$$

$$= \boxed{3\frac{6}{7}}$$

Comparing these two methods, notice that when cancellation of factors was not used, the <u>fives</u> were still divided out, but at a <u>later</u> point in the problem. The cancellation method combined both the multiplication and division steps.

---

*Refer to Section 3.4 for a discussion of factors

**Refer to Section 3.3 for a discussion of reducing fractions.

Example 3: (Multiplication of Mixed Numbers)
❏ Multiply 5-5/8 x 2-1/5.

First, convert mixed numbers to fractions:

$$5\frac{5}{8} \ = \ \boxed{\frac{45}{8}} \qquad 2\frac{1}{5} \ = \ \boxed{\frac{11}{5}}$$

The problem can now be written as shown below. First, look for common factors in the numerators and denominators:

$$\frac{\overset{9}{\cancel{45}}}{8} \times \frac{11}{\underset{1}{\cancel{5}}} \ = \ \frac{99}{8}$$

$$= \ \boxed{12\frac{3}{8}}$$

MULTIPLYING MIXED NUMBERS

When multiplying mixed numbers:

1. **Convert** mixed numbers to improper fractions*.

2. **Continue** as for fractions.

Example 4: (Multiplication of Fractions)
❏ Multiply 4 x 6-1/3.

$$4 \times 6\frac{1}{3}$$

Write both of these numbers as fractions:

$$\frac{4}{1} \times \frac{19}{3} \ = \ \frac{76}{3}$$

$$= \ \boxed{25\frac{1}{3}}$$

When you wish to write a whole number as a fraction, simply make the denominator 1. For example, 4 = 4/1, 27 = 27/1, etc.

Example 5: (Multiplication of Fractions)
❏ Multiply 5-1/4 x 3 x 2-2/3.

Convert all three numbers to fractions:

$$\frac{21}{4} \times \frac{3}{1} \times \frac{8}{3} \ =$$

Check for any factors which can be divided out, then multiply numerators and multiply denominators:

$$\frac{\overset{7}{\cancel{21}}}{\underset{1}{\cancel{4}}} \times \frac{3}{1} \times \frac{\overset{2}{\cancel{8}}}{\underset{1}{\cancel{3}}} \ = \ \frac{42}{1}$$

$$= \ \boxed{42}$$

More than two fractions or mixed numbers may be multiplied together. Use the same steps in multiplying.

*Refer to Section 3.5 for discussion of mixed numbers to improper fractions.

PRACTICE PROBLEMS 3.7: Mutiplication of Fractions and Mixed Numbers

❑ Multiply as indicated, using cancellation of common factors when possible, and reduce your answers to lowest terms.

1. $\dfrac{2}{3} \times \dfrac{1}{3} =$ _____

2. $\dfrac{7}{8} \times \dfrac{11}{49} =$ _____

3. $\dfrac{15}{16} \times \dfrac{48}{6} =$ _____

4. $1\dfrac{2}{3} \times \dfrac{3}{4} =$ _____

5. $\dfrac{3}{3} \times 15 =$ _____

6. $\dfrac{1}{7}$ of $2\dfrac{1}{2} =$ _____

7. $8 \times 7\dfrac{9}{12} =$ _____

8. $9\dfrac{3}{8} \times 5 =$ _____

9. $\dfrac{5}{6} \times \dfrac{9}{5} =$ _____

10. $\dfrac{7}{8} \times 10 \times \dfrac{4}{21} \times \dfrac{2}{15} =$ _____

3.8 DIVISION BY FRACTIONS AND MIXED NUMBERS

<table>
<tr><td>

SUMMARY

To divide by a **fraction**:

• **Invert** the denominator.

• **Multiply** fractions.

$$3 \div \frac{1}{2} =$$

"How many one-halves are there in the number three?"

Invert the denominator and multiply:

$$3 \div \frac{1}{2} = \frac{3}{1} \times \frac{2}{1}$$

$$= \boxed{6}$$

To divide by a **mixed number**:

• **Convert** the mixed number to an improper fraction.

• **Invert** the denominator.

• **Multiply** fractions.

</td></tr>
</table>

To understand division by fractions, let's first review division by whole numbers. Given the problem 30 ÷ 5, what you are really saying is, "how many 5's are there in the number 30?" (The answer is 6.) Similarly, 56 ÷ 7 really means, "how many groups of 7 are there in the number 56?" (The answer is 8.)

When dividing by fractions, the same type question can be posed. For example, 3 ÷ 1/2 means, "how many one-halves are there in the number 3?" (The answer is 6.)

DIVISION BY FRACTIONS

To divide by fractions:

1. **Invert the denominator.** To invert a number, you flip it over—what was on top goes on the bottom, and vice versa.

2. **Multiply** fractions. (Don't forget to divide out common factors before multiplying.)

Example 1: (Division by Fractions)
❑ Complete the following: 3/5 ÷ 1/4.

$$\frac{3}{5} \div \frac{1}{4} \quad \text{can also be written as} \quad \frac{\frac{3}{5}}{\frac{1}{4}}$$

Note that the term <u>following</u> the division sign is in the denominator.

Invert the denominator and multiply:

$$\frac{3}{5} \div \frac{1}{4} \longrightarrow \frac{3}{5} \times \boxed{\frac{4}{1}} \longleftarrow \text{Inverted}$$

$$\frac{3}{5} \times \frac{4}{1} = \frac{12}{5}$$

$$= \boxed{2\frac{2}{5}}$$

Example 2: (Division by Fractions)
❑ Divide as indicated: 3/8 ÷ 24/15.

$$\frac{3}{8} \div \frac{24}{15} \longleftarrow \text{Invert the denominator and multiply:}$$

$$\frac{\overset{1}{\cancel{3}}}{8} \times \frac{15}{\underset{8}{\cancel{24}}} = \boxed{\frac{15}{64}}$$

When a whole number is in the numerator or denominator, write it as a fraction by putting the whole number over 1. For example, 7 = 7/1 and 5 = 5/1. Then continue as usual.

Example 3: (Division by Fractions)
❑ Complete the following: 15 ÷ 1/3.

$$15 \div \frac{1}{3} = \frac{15}{1} \times \frac{3}{1}$$

$$= \frac{45}{1}$$

$$= \boxed{45}$$

Example 4: (Division with Mixed Numbers)
❑ Divide as follows: 2-1/5 ÷ 1-2/3.

First, convert the mixed numbers to fractions:

$$2\frac{1}{5} \div 1\frac{2}{3} = \frac{11}{5} \div \frac{5}{3}$$

Then invert and multiply:

$$= \frac{11}{5} \times \frac{3}{5}$$

$$= \frac{33}{25}$$

$$\text{or} = \boxed{1\frac{8}{25}}$$

DIVISION WITH MIXED NUMBERS

Division with mixed numbers is the same as division by fractions except for the first step:

1. **Convert** the mixed number to an improper fraction.

2. **Invert** the denominator.

3. **Multiply** fractions.

Example 5: (Division with Mixed Numbers)
❑ Divide: 20 ÷ 15-1/4.

Convert the whole number and mixed number to fractions:

$$20 \div 15\frac{1}{4} = \frac{20}{1} \div \frac{61}{4}$$

Then invert and multiply:

$$= \frac{20}{1} \times \frac{4}{61}$$

$$= \frac{80}{61}$$

$$\text{or} = \boxed{1\frac{19}{61}}$$

Example 6: (Division with Mixed Numbers)
❑ Divide as indicated: 5/9 ÷ 10/45.

$$\frac{5}{9} \div \frac{10}{45} = \frac{\overset{1}{\cancel{5}}}{\underset{1}{\cancel{9}}} \times \frac{\overset{5}{\cancel{45}}}{\underset{2}{\cancel{10}}}$$

$$= \frac{5}{2} \text{ or } \boxed{2\frac{1}{2}}$$

PRACTICE PROBLEMS 3.8: Division by Fractions and Mixed Numbers

❏ Complete the following problems. Reduce your answer to lowest terms.

1. $\dfrac{3}{5} \div \dfrac{10}{4} =$ _____

2. $4 \div 11\dfrac{2}{9} =$ _____

3. $\dfrac{7}{8} \div \dfrac{14}{24} =$ _____

4. $\dfrac{2}{3} \div \dfrac{1}{5} =$ _____

5. $8\dfrac{3}{7} \div 5 =$ _____

6. $\dfrac{1}{5} \div 2\dfrac{5}{13} =$ _____

7. $1\dfrac{1}{8} \div \dfrac{5}{16} =$ _____

8. $\dfrac{7}{16} \div \dfrac{7}{8} =$ _____

9. $\dfrac{2}{7} \div \dfrac{16}{21} =$ _____

10. $12\dfrac{4}{5} \div 3\dfrac{3}{8} =$ _____

3.9 COMBINED CALCULATIONS WITH FRACTIONS

SUMMARY

A **complex fraction** is a fraction whose numerator and/or denominator is also a fraction. Examples of complex fractions include:

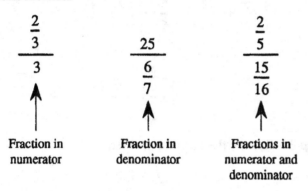

$$\frac{\frac{2}{3}}{3} \qquad \frac{25}{\frac{6}{7}} \qquad \frac{\frac{2}{5}}{\frac{15}{16}}$$

Fraction in numerator   Fraction in denominator   Fractions in numerator and denominator

Occasionally you will come across a problem that combines complex fractions in several mathematical operations, such as:

$$\frac{\frac{1}{8} + \frac{7}{16}}{\frac{2}{5} + \frac{1}{3}} \qquad \text{or} \qquad \frac{2\frac{7}{8} + \frac{1}{5}}{10 \div \frac{3}{4}}$$

To complete these problems:

- **Simplify** the numerator and denominator.

- **Restate** the original problem.

- **Divide** as indicated by the restated problem.

Problems involving combined calculations can involve addition, subtraction, multiplication, and division of fractions all in the same problem. Therefore, it is advisable that you be comfortable with the calculations given in Sections 3.6 - 3.8 before attempting this section.

COMBINED CALCULATIONS WITH FRACTIONS

Combined calculations are simply several calculations in one problem. Simplifying the numerator represents one calculation, simplifying the denominator represents another calculation, and completing the final calculation after restating the problem represents a third calculation—all in the same problem. Following the same procedure each time you do a combined calculation will help eliminate errors:

1. **Simplify** the numerator and denominator by completing the calculation indicated.

2. **Restate** the original problem using results from Step 1.

3. **Divide** as indicated by the restated problem.

Example 1:   (Combined Calculations)
❑ Complete the following problem:

$$\frac{\frac{5}{6} + 2\frac{1}{3}}{5}$$

Step 1—Simplify the numerator of this problem:

$$\frac{5}{6} + 2\frac{1}{3} = \frac{5}{6} + 2\frac{2}{6}$$
$$= 2\frac{7}{6}$$
$$= \boxed{3\frac{1}{6}}$$

Step 2—Restate the original problem:

$$\frac{3\frac{1}{6}}{5}$$

Step 3—Divide as indicated:

$$3\frac{1}{6} \div \frac{5}{1} = 3\frac{1}{6} \times \frac{1}{5}$$
$$= \frac{19}{6} \times \frac{1}{5}$$
$$= \boxed{\frac{19}{30}}$$

Example 2:   (Combined Calculations)
❑ Calculate as indicated:

$$\frac{\frac{1}{3} \times 1\frac{1}{2} \times 5}{7 \div \frac{5}{8}}$$

Step 1—Simplify numerator and denominator:

Numerator: $\dfrac{1}{\cancel{3}_1} \times \dfrac{\cancel{3}^1}{2} \times \dfrac{5}{1} = \dfrac{5}{2}$

Continued on next page...

Example 2 Continued:

Denominator: $\dfrac{7}{1} \div \dfrac{5}{8} = \dfrac{7}{1} \times \dfrac{8}{5}$

$$= \dfrac{56}{5}$$

Step 2—Restate the original problem:

$$\dfrac{\dfrac{5}{2}}{\dfrac{56}{5}}$$

Step 3—Divide as indicated:

$$\dfrac{5}{2} \div \dfrac{56}{5} = \dfrac{5}{2} \times \dfrac{5}{56}$$

$$= \boxed{\dfrac{25}{112}}$$

Example 3: (Combined Calculations)
❑ Solve the problem shown below.

$$\dfrac{\dfrac{3}{10} + 1\dfrac{2}{5} + \dfrac{12}{25}}{5 - \dfrac{7}{8} + \dfrac{1}{4}}$$

Simplify numerator and denominator, restate the problem, then divide as indicated:

$$\dfrac{\dfrac{109}{50}}{\dfrac{35}{8}} = \dfrac{109}{50} \div \dfrac{35}{8}$$

$$= \dfrac{109}{\underset{25}{\cancel{50}}} \times \dfrac{\overset{4}{\cancel{8}}}{35}$$

$$= \boxed{\dfrac{436}{875}}$$

PRACTICE PROBLEMS 3.9: Combined Calculations With Fractions

❑ Complete the problems shown below. Reduce your answer to lowest terms.

1. $\dfrac{2\frac{1}{6} + 3\frac{7}{8}}{4\frac{1}{3}} =$ _____

2. $\dfrac{\frac{3}{4} + \frac{4}{7}}{\frac{1}{4} + \frac{5}{9}} =$ _____

3. $\dfrac{\frac{1}{3} \text{ of } 10}{\frac{7}{12}} =$ _____

4. $\dfrac{\frac{2}{3} \times 1\frac{1}{2}}{5} =$ _____

5. $\dfrac{\frac{1}{5} + \frac{3}{4}}{\frac{9}{10} - \frac{1}{3}} =$ _____

3.10 DECIMAL FRACTIONS

SUMMARY ONLY

Fractions that have a denominator of 10, or 10 with one or more zeros after it (10, 100, 1000, etc.), can be written in a different way using the decimal system. Several common fractions and their equivalent decimal fractions are shown below.

COMMON FRACTIONS & DECIMAL FRACTIONS

$$1/10 \longrightarrow 0.1$$

$$5/100 \longrightarrow 0.05$$

$$4/1000 \longrightarrow 0.004$$

$$17/10,000 \longrightarrow 0.0017$$

$$295/100,000 \longrightarrow 0.00295$$

A detailed discussion of the decimal system and how to convert from fractions to decimals, and vice versa, is given in Chapter 4, Decimals.

3.11 OTHER MEANINGS OF FRACTIONS

SUMMARY ONLY

In this chapter we have focused on the concept of a fraction as a "part of a whole." However, there are other meanings of fractions which are equally important.

Fractions can represent **a relationship of one number to another,** such as fractions that represent **ratios** (1 gallon of chemical mixed with 15 gallons of water is a 1/15 ratio), or fractions that represent **rates** (a flow of 3 cubic feet per minute can be written 3 cu ft/1 min.) Ratios and proportions are discussed in Chapter 7. Flow rate calculations are discussed in Chapter 8.

Fractions can also represent percents. Calculations of percent are discussed in Chapter 5.

NOTES:

4 *Decimals*

4.1 The Decimal System

Number Correct

❏ Write a fraction and decimal number for the shaded area of each figure shown below. Reduce the fraction, if possible.

1. 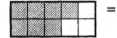 = _____ _____

4. 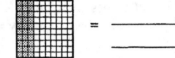 = _____ _____

2. = _____ _____

5. 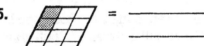 = _____ _____

3.  = _____ _____

❏ Write a mixed number and decimal number for the shaded area of each figure shown below. Reduce the fraction, if possible.

6.   = _____ _____

7. = _____ _____

8. = _____ _____

9. = _____ _____

10. = _____ _____

4.2 Addition and Subtraction of Decimals

❑ Complete the following addition or subtraction problems.

1. $16.5 + 45.8$ = _____

2. $0.59 + 0.7$ = _____

3. $125.14 - 31.9$ = _____

4. $40.5 - 18.125$ = _____

5. $421 - 24.55$ = _____

6. $0.5 + 7.53 + 9.12$ = _____

7. $1.03 - 0.323$ = _____

8. $115.5 + 17.2 - 25.25$ = _____

❑ Complete the following word problems.

9. What was the total flow for a week (in MG), given the following information: Monday = 3.6 MG; Tuesday = 3.41 MG; Wednesday = 3.25 MG; Thursday = 3.71 MG; Friday = 3.84 MG; Saturday = 3.92 MG; Sunday = 3.32 MG.

ANS_____

10. A 100-m*L* sludge sample has been dried. The dish weighs 17.2500 grams. The dish and dried solids weigh 17.2608 grams. What is the weight of the dried solids only?

ANS_____

4.3 Multiplication of Decimals

❑ In the following problems, multiply as indicated.

1. 125.9×14.52 = _____

2. 0.15×0.003 = _____

3. 3×24.7291 = _____

4. $(3.14)(90)$ = _____

5. 1.25×72.15 = _____

6. $25 \times \$82.71$ = _____

7. $(0.785)(25)(25)$ = _____

8. $(2.31)(7.48)(8.34)$ = _____

❑ Complete the following word problems.

9. How many pounds of water will a 55-gallon barrel hold?
(To convert from gallons to pounds, multiply by 8.34.)   ANS_____

10. 28 psi is equivalent to how many feet of head?
(To convert from psi to ft of head, multiply by 2.31.)   ANS_____

4.4 Division of Decimals

1. $42.7 + 20 =$ _____

2. $212.25 + 25 =$ _____

3. $6794 + 0.22 =$ _____

4. $157.25 + 7.48 =$ _____

5. $13.67 + 3 =$ _____

6. $4369.5 + 2.5 =$ _____

7. $12,728 + 8.34 =$ _____

8. $0.85 + 0.22 =$ _____

❏ Complete the following word problems.

9. A flow of 708 gpm is equivalent to how many cfm?
   (To convert from gpm to cfm, divide gpm by 7.48.)   ANS_____

10. A tank contains 35,000 lbs of water. How many gallons is this?
   (To convert from lbs to gal, divide the lbs by 8.34.)
   ANS_____

4.5 Converting Decimals and Fractions

❏ Convert the following decimals to fractions or mixed numbers. (For these problems, do not reduce fractions to lowest terms.)

1. $7.5 =$ _____

2. $0.92 =$ _____

3. $0.005 =$ _____

4. $12.75 =$ _____

5. $627.8 =$ _____

❏ Convert the following fractions or mixed numbers to decimals.

6. $15\dfrac{1}{3} =$ _____

7. $2\dfrac{7}{100} =$ _____

8. $2140\dfrac{1}{8} =$ _____

9. $84\dfrac{6}{7} =$ _____

10. $\dfrac{1}{290} =$ _____

NOTES:

4.1 THE DECIMAL SYSTEM

Calculations involving decimal numbers will be much easier if you understand the basic concepts underlying the decimal system. The word **decimal** comes form the Latin word *decem*, meaning ten; thus the decimal system is based on **ten and multiples of ten** (sometimes called "powers" of ten*). The postition of the decimal point and the place values are of great importance in this system.

> ### SUMMARY
>
> 1. The value of any number in the decimal system depends on the placement of the decimal point.
>
> 2. Common fractions and decimal fractions are two ways to express the same quantity:
>
>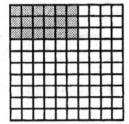
>
> | Common Fraction | Decimal Fraction |
> |:---:|:---:|
> | $\dfrac{18}{100}$ | 0.18 |
>
> 3. Mixed numbers and decimal numbers express numbers greater than one:
>
>
>
> | Mixed Number | Decimal Number |
> |:---:|:---:|
> | $3\dfrac{1}{10}$ | 3.1 |

---

*"Powers of ten" refers to ten multiplied by itself a designated number of times. The exponent tells us how many times 10 should be multiplied. Thus, $10^2 = (10)(10)$; $10^3 = (10)(10)(10)$; etc.

PLACE VALUE SYSTEM

In a place value system, the value (or size) of any number depends on two things:

1. Which digits are used (0,1,2...9), and

2. Where these digits are placed in relation to the decimal point.

For example, using the same digits 1725, you can write many different numbers (172.5, 1.725, 0.01725, etc.), each with a very different value.

The **value** of any number in the decimal system, therefore **depends on the placement of the decimal point**. The numbers to the left of the decimal point represent whole numbers (any number equal to or greater than one). The numbers to the right of the decimal point represent numbers less than one.

WHOLE NUMBERS

The place values of whole numbers are shown in the diagram to the right. When no decimal is shown for a number, **the decimal point is assumed to be at the far right.** For example, 526 is assumed to be 526.0.

DECIMAL FRACTIONS

Just as proper fractions* represent portions or quantities less than one, the numbers to the right of the decimal point represent quantities less than one. These numbers are called decimal fractions. As with whole numbers the value of the decimal fraction depends on which place values are used.

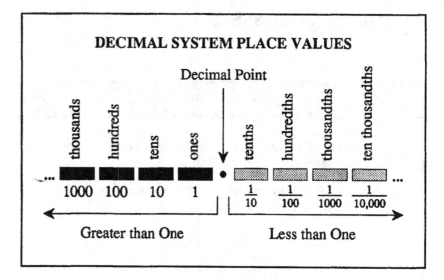

DECIMAL SYSTEM PLACE VALUES

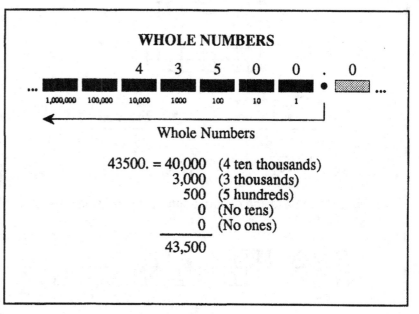

WHOLE NUMBERS

$$43500. = 40,000 \quad \text{(4 ten thousands)}$$
$$3,000 \quad \text{(3 thousands)}$$
$$500 \quad \text{(5 hundreds)}$$
$$0 \quad \text{(No tens)}$$
$$0 \quad \text{(No ones)}$$
$$\overline{43,500}$$

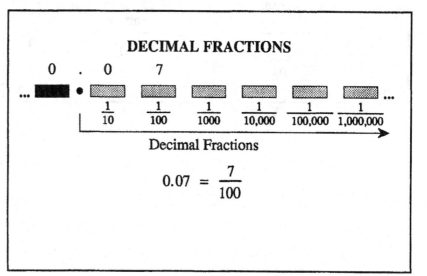

DECIMAL FRACTIONS

$$0.07 = \frac{7}{100}$$

*A "proper fraction" is a fraction whose numerator is **smaller** than its denominator.

Example 1: (Writing Decimal Fractions)
❑ Write a common fraction and decimal fraction for the shaded area shown below.

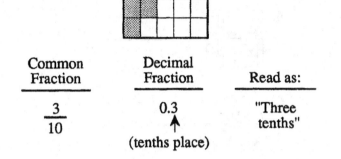

Common Fraction	Decimal Fraction	Read as:
$\frac{3}{10}$	0.3 ↑ (tenths place)	"Three tenths"

Example 2: (Writing Decimal Fractions)
❑ Write a decimal fraction for the shaded area shown below.

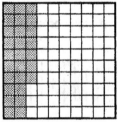

Common Fraction	Decimal Fraction	Read as:
$\frac{27}{100}$	0.27 ↑ (hundredths place)	"Twenty-seven hundredths"

Example 3: (Writing Decimal Fractions)
❑ Write a mixed number and decimal number for the shaded area shown below.

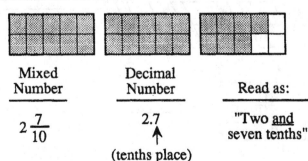

Mixed Number	Decimal Number	Read as:
$2\frac{7}{10}$	2.7 ↑ (tenths place)	"Two <u>and</u> seven tenths"

READING AND WRITING DECIMAL FRACTIONS

Decimal fractions are closely related to common or regular fractions. In fact, decimal fractions are simply a different way of writing common fractions which have a denominator of 10, 100, 1000, 10,000, etc.

The common fraction for the shaded area shown in Example 1 is 3/10. To write a decimal fraction for the same shaded area, count the number of shaded rectangles and put that number in the place value position for tenths (the figure has been divided into tenths): 0.3.

The common and decimal fractions in this example would be read as *three tenths*. Notice the "th" sound is given to all the place values to the right of the decimal point.

In Example 2, the common fraction for the shaded area is 27/100. To write a decimal fraction for the same area, the 7 in the number 27 must be placed in the hundredths place value, as shown. This decimal fraction would be read as *twenty-seven hundredths*.

READING AND WRITING DECIMAL NUMBERS

To write a number which includes both whole numbers and fractions, you can use a mixed number, such as 32-17/100 or a decimal number, 32.17. When writing a decimal number, the "wholes" are written to the left of the decimal point and the fractional part to the right of the decimal point. Notice that the decimal point is read as *and*.

PRACTICE PROBLEMS 4.1: The Decimal System

❏ Write a common fraction <u>and</u> decimal fraction for the shaded areas of each figure below.

1. 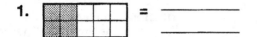 = _____

4. 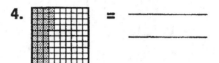 = _____

2. = _____

5. 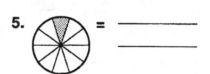 = _____

3. 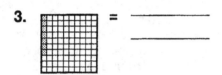 = _____

❏ Write a mixed number <u>and</u> decimal number for the shaded areas of each figure below. (Do not reduce the fractions.)

6. 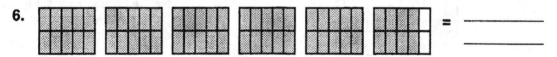 = _____

7. 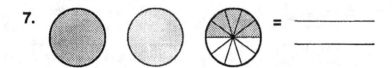 = _____

8.  = _____

9.  = _____

10. 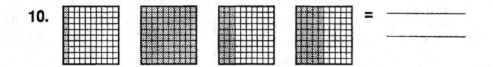 = _____

4.2 ADDITION AND SUBTRACTION OF DECIMALS

SUMMARY

- When adding or subtracting decimal numbers, the fundamental concept to remember is that **the decimal points must be aligned**.

- If no decimal point is shown for a number, place a decimal point to the right of the number (e.g. 18 = 18.)

- For better column alignment, zeros are sometimes added to the right of a number.

- Add or subtract as with whole numbers.

Addition and subtraction of decimals is most often done using a calculator. However, situations arise when calculations must be done by hand. Decimal addition and subtraction is perhaps most commonly required in determining averages and in making laboratory calculations used to determine the amount of suspended and volatile solids in a sample.

ADDING AND SUBTRACTING DECIMAL NUMBERS

Adding and subtracting decimal numbers is like adding and subtracting whole numbers, with one exception: **the decimal points must align.** This is so that you will be adding or subtracting the same place values—tenths with tenths, hundredths with hundredths, and so on.

When a number does not have a decimal point shown, place a decimal point to the right of the last digit of that number.

Aligning numbers is sometimes easier if you **add zeros to the right of the numbers** so that all the columns line up evenly. This does not change the value of the numbers, but it helps eliminate errors in addition and subtraction problems.

Example 1: (Adding Decimals)
❑ Add the following numbers:

$$
\begin{array}{r}
3.41 \\
0.97 \\
+\ 0.125 \\
\hline
4.505 \\
\end{array}
$$

↑

Align Decimal Points

Example 2: (Adding Decimals)
❑ Add 27.5, 15 and 16.3

$$
\begin{array}{r}
27.5 \\
15. \\
+\ 16.3 \\
\hline
58.8 \\
\end{array}
$$

Add a decimal point to the **right** when one is not shown.

Example 3: (Adding Decimals)
❑ Add 0.64, 481.1 and 0.005.

$$
\begin{array}{r}
0.64\boxed{0} \\
481.1\boxed{00} \\
+\ 0.005 \\
\hline
481.745 \\
\end{array}
$$

Add zeros to the right to help align columns.

Example 4: (Subtracting Decimals)
❑ Complete the following: 15.816 - 7.72.

$$
\begin{array}{r}
15.816 \\
- \ 7.72\boxed{0} \\
\hline
8.096
\end{array}
$$

↑
Align Decimal Points

Example 5: (Subtracting Decimals)
❑ Subtract 0.45 from 19.

$$
\begin{array}{r}
19.\boxed{00} \\
- \ 0.45 \\
\hline
18.55
\end{array}
$$

← Add a decimal point, then two zeros.

Example 6: (Subtracting Decimals)
❑ What is 453.2 minus 54.375?

$$
\begin{array}{r}
453.200 \\
- \ 54.375 \\
\hline
398.825
\end{array}
$$

Adding zeros to align the columns is particularly important for subtraction problems since borrowing is often required.

PRACTICE PROBLEMS 4.2: Addition and Subtraction of Decimals

❏ Complete the following addition or subtraction problems.

1. $6.9 + 5.7 + 4 =$ _____

2. $0.1654 + 3.139 =$ _____

3. $18.1 - 7.725 =$ _____

4. $0.7931 - 0.3989 =$ _____

5. $6.815 - 5.931 =$ _____

6. $63 - 28.57 =$ _____

7. $6.09 + 3.23 + 9.712 =$ _____

8. $5.9 + 7.62 + 4 + 6.93 =$ _____

❏ Complete the following word problems.

9. Given the following daily flow information, what has the total flow, in MG, for the week? Monday = 5.92 MG; Tuesday = 5.1 MG; Wednesday = 4.7 MG, Thursday = 5.22 MG; Friday = 4.7 MG; Saturday = 5.3 MG; Sunday = 4.45 MG.

ANS_____

10. For a suspended solids test, the amount of suspended solids in water is measured by the solids remaining in a dish after drying overnight. If the weight of the dish is 24.1982 grams, and the weight of the dish and dried solids is 24.2045, what is the weight (in grams) of the dried solids alone?

ANS_____

4.3 MULTIPLICATION OF DECIMALS

SUMMARY
When mulitiplying decimal numbers: • Multiply the numbers, disregarding decimal points. • Considering all the numbers being multiplied, find the **total number of places to the right of the decimal point.** • Place the decimal point in the answer.

Multiplication of decimal numbers is an essential calculation in water and wastewater math. It is used in making volume conversions (cubic feet to gallons), concentration conversions (milligrams per liter to pounds per day), and pressure conversions (feet of head to pounds per square inch). A weakness in this area of basic math would very severely hamper your ability to succeed in water and wastewater calculations.

MULTIPLY THEN COUNT DECIMAL PLACES

Multiplying decimal numbers is similar to multiplying whole numbers. The only difference is that with decimal numbers **you must consider decimal places to the right of the decimal point:**

1. Multiply the numbers given, disregarding any decimal points.

2. Count the decimal places to the right of the decimal point for each number being multiplied.

3. Place the decimal point in the answer. Your answer should have the total number of decimal places to the right of the decimal point, as counted in Step 2.

Example 1: (Multiplying Decimals)
❏ Multiply 19.5 by 16.5.

$$
\begin{array}{r}
19.5 \\
\times\ 16.5 \\
\hline
975 \\
1170 \\
1950 \\
\hline
321.75
\end{array}
$$

19.5 ◄— There is one decimal place to the right of the decimal point.

× 16.5 ◄— There is one decimal place to the right of the decimal point.

Total = <u>2 decimal places</u> to the right of the decimal point.

Place the decimal point so there are <u>2 decimal places</u> to the right of the decimal point.

Example 2: (Multiplying Decimals)
❏ Find the product of 0.47 and 0.7.

$$
\begin{array}{r}
0.47 \\
\times\ 0.7 \\
\hline
.329
\end{array}
$$

0.47 ◄— There are two decimal places to the right of the decimal point.

× 0.7 ◄— There is one decimal place to the right of the decimal point.

Total = <u>3 decimal places</u> to the right of the decimal point.

Place the decimal point in the answer so there are <u>3 decimal places</u> to the right of the decimal point.

Example 3: (Multiplying Decimals)
❏ Multiply 0.028 by 0.05.

$$0.028 \longleftarrow$$ There are <u>three decimal places</u> to the right of the decimal point.

$$\times \; 00.05 \longleftarrow$$ There are <u>two decimal places</u> to the right of the decimal point.

$$.00140$$

Two zeros added $\longrightarrow$

Place the decimal point in the answer so there are <u>5 decimal places</u> to the right of the decimal point. (Two zeros had to be added in order to place the decimal point in the proper position.)

ADDING ZEROS

Sometimes it is necessary to add zeros in order to place the decimal point in the answer.

Example 4: (Multiplying Decimals)
❏ Find the product of 0.33 and 10.

$$0.33 \times 10$$

<u>One zero</u> in the multiplier = <u>move the decimal point one place</u> to the right.

$$0.33 = \boxed{3.3}$$

MULTIPLYING BY TEN OR A MULTIPLE OF TEN*

When multiplying a decimal number by 10, 100, 1000, etc., there is a **shortcut** that can be used—count the number of zeros in the multiplier, and **move the decimal point to the right that many places.**

Example 5: (Multiplying Decimals)
❏ Multiply 95.5 by 100.

$$95.5 \times 100$$

<u>Two zeros</u> in multiplier = move decimal point <u>two places</u> to the right.

$$95.50 = \boxed{9550.}$$

Hint: If you forget whether to move the decimal point to the right or left, remember that:

• Multiplying by whole numbers results in a larger number, and

•To obtain a larger number, you must move the decimal point to the right.

*Refer to Chapter 13 for a discussion of powers of ten.

PRACTICE PROBLEMS 4.3: Multiplication of Decimals

❑ Multiply the following decimal numbers as indicated.

1. 17.4 x 98.26 = _____

5. 2.87 x 0.51 = _____

2. $45.23 x 15 = _____

6. 215.09 x 0.375 = _____

3. (0.785)(30)(30) = _____

7. 6.5 x 1.3 x 2.5 = _____

4. 3 x 0.5488 = _____

8. (0.52)(1.7)(1.7)(0.785) = _____

❑ Complete the following word problems.

9. The capacity of a tank is 10,000 cu ft. How many gallons will the tank hold? (To convert from cubic feet volume to gallons volume, multiply the number of cubic feet by 7.48)

ANS_____

10. Suppose the suspended solids concentration of water entering a clarifier is 110 mg/L. If the flow to the clarifier is 3 MGD, how many lbs/day suspended solids are entering the clarifier? (To convert from mg/L to lbs/day concentration, multiply as follows: (mg/L)(flow, MGD)(8.34) = lbs/day.)

ANS_____

4.4 DIVISION OF DECIMALS

SUMMARY

1. **To divide decimals by whole numbers:**

   - Place the decimal point in the answer (directly above the number being divided).

   - Divide as with whole numbers.

2. **To divide decimals by decimals:**

   - Move the decimal point in the divisor to the far right.

   - Move the decimal point in the dividend (the number under the division sign) the same number of places to the right.

   - Place the decimal point in the answer.

   - Divide as with whole numbers.

Division involving decimal numbers is also a key skill in water and wastewater math. Perhaps as much as 80% of the calculations involve either multiplication or division of decimals, or both.

DIVIDING DECIMALS BY WHOLE NUMBERS

To divide a decimal number by a whole number:

1. Place the decimal point in the answer (quotient) directly above the decimal point in the number being divided (dividend).

2. Divide as with whole numbers rounding* your answer as required.

DIVIDING DECIMALS BY DECIMALS

When dividing a decimal number by another decimal number you must first reposition decimal points**:

1. Move the decimal point in the divisor (the number you are dividing by) to the far right. This makes the divisor a whole number.

2. Move the decimal point in the dividend (the number under the division sign) the same number of places to the right.

   After moving the decimal points, continue as you would for dividing a decimal by a whole number (see Example 1):

3. Place a decimal point in the answer.

4. Divide as with whole numbers, rounding as required.

Example 1: (Dividing Decimals)
❑ Divide 145.7 by 3. Give your answer to the nearest tenth.

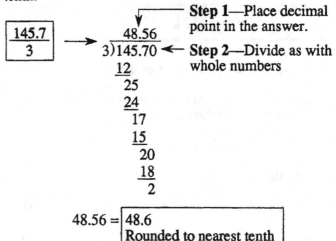

Step 1—Place decimal point in the answer.

Step 2—Divide as with whole numbers

48.56 = 48.6
Rounded to nearest tenth

Example 2: (Dividing Decimals)
❑ Complete the following: 20.9 ÷ 0.55. Round your answer to the nearest hundredth.

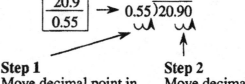

Step 1
Move decimal point in divisor to far right.

Step 2
Move decimal point in dividend same number of places to right.

Continue as dividing a decimal by a whole number:

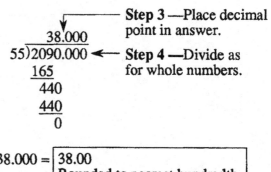

Step 3 —Place decimal point in answer.

Step 4 —Divide as for whole numbers.

38.000 = 38.00
Rounded to nearest hundredth

---

*Refer to Chapter 14, Rounding and Estimating.

**Remember, moving the decimal point to the right is the same as multiplying by ten or a multiple of ten. Since you are moving the **decimal point in both numbers of the division problem the same number of places to the right, you have not changed the value of the **division problem. For example, 3/7 = 30/70. The value remains the same.

Example 3: (Dividing by Decimals)
❑ Divide 45 by 0.72. Round your answer to the nearest whole number.

$$\frac{45}{0.72} \longrightarrow 0.72\overline{)45.00}$$

Step 1   Step 2

Step 3

$$\begin{array}{r} 62.50 \\ 72\overline{)4500.00} \\ \underline{432} \\ 180 \\ \underline{144} \\ 360 \\ \underline{360} \\ 0 \end{array}$$

Step 4

$62.50 = \boxed{\begin{array}{l} 62.5 \\ \text{Rounded to nearest tenth} \end{array}}$

Example 4: (Dividing by Decimals)
❑ Complete the following: 1865.9 ÷ 100.

$$1865.9 \div 100$$

Two zeros in the divisor = move the decimal point two places to the left.

$$1865.9 = \boxed{18.659}$$

Example 5: (Dividing by Decimals)
❑ Divide 85.7 by 1000

$$85.7 \div 1000$$

Three zeros in divisor = move the decimal point three places to the left.

$$085.7 = \boxed{0.0857}$$

DIVIDING BY TEN OR MULTIPLE OF TEN*

When dividing a decimal number by 10, 100, 1000, etc., there is a shortcut that can be used—count the number of zeros in the divisor, and move the decimal point to the left that many places.

Hint: If you forget which direction to move the decimal point when dividing, remember that:

- Dividing by whole numbers results in a smaller number, and

- To obtain a smaller number, you must move the decimal point to the left.

Sometimes it will be necessary to add zeros in order to place the decimal point, as shown in Example 5.

* Refer to Chapter 13 for a review of powers of ten.

PRACTICE PROBLEMS 4.4: Division of Decimals

❑ Complete the following division problems.

1. $87.62 \div 20 =$ _____

2. $10.9 \div 15 =$ _____

3. $900 \div 0.785 =$ _____

4. $0.9 \div 0.433 =$ _____

5. $425 \div 0.22 =$ _____

6. $17.9 \div 8.34 =$ _____

7. $6.018 \div 1.82 =$ _____

8. $42.500 \div 7.48 =$ _____

❑ Complete the following word problems.

9. 50 lbs of water are equivalent to how many gallons?
(To convert from lbs to gal, divide the number of lbs by 8.34)

ANS_____

10. 2500 lbs of suspended solids enter the sedimentation tank each day. If the average daily suspended solids contributed by each person is 0.22 lbs of suspended solids, what size population is represented by the 2500 lbs suspended solids load? (To find out how many 0.22 lbs are in 2500, divide 2500 lbs by 0.22 lbs.)

ANS_____

4.5 CONVERTING DECIMALS AND FRACTIONS

Decimals and fractions are two ways of expressing a quantity. Fractions may sometimes be preferred. For example, the fraction 1/3 is a more **precise** description of a quantity than 0.33, which must be rounded. For convenience and ease in working with other numbers, decimals are often preferred. It is important, therefore, to be able to convert fractions to decimals and vice versa.

SUMMARY

1. **To convert a decimal to a fraction:**

   - The entire number, disregarding the decimal point, becomes the **numerator of the fraction.**
   - The place value of the last number to the right indicates the **denominator of the fraction.**

$$0.53 = \frac{53}{100}$$

   - Reduce the fraction if possible.

2. **To convert a decimal to a mixed number:**

   - Write the whole number (the number to the <u>left</u> of the decimal point).
   - The number to the <u>right</u> of the decimal point is the numerator of the fraction.
   - The place value of the last number to the <u>right</u> indicates the denominator of the fraction.
   - Reduce the fraction if possible.

$$14.9 = 14\frac{9}{10}$$

3. **To convert a fraction to a decimal:**

   - Write the fraction as a division problem and continue as described in Section 4.4.

$$\frac{4}{7} = 7\overline{)4.000} \begin{array}{r} .57 \\ \hline \end{array}$$

$$\begin{array}{r} \underline{35} \\ 50 \\ \underline{49} \\ 1 \end{array}$$

4. **To convert a mixed number to a decimal:**

   - Write the whole number and place a decimal number to the right of it.
   - Convert the fractional portion of the mixed number to a decimal, as described above.

$$12\frac{1}{5} = 12.2$$

CONVERTING DECIMALS TO FRACTIONS

Whether you wish to convert decimals less than one (e.g., 0.52), or greater than one (e.g., 14.9), the steps in converting to fractions are the same:

1. The entire number, regardless of the decimal point, becomes the numerator of the fraction. (Disregard any zeros to the left or right of the number. Zeros between other numbers must be retained.)

2. The position (place value)* of the last number to the right of the decimal point indicates the denominator of the fraction.

3. Reduce the fraction, if possible.**

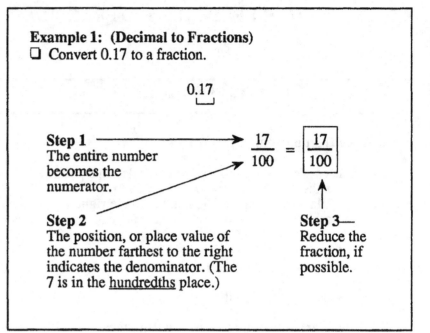

Example 1: (Decimal to Fractions)
❑ Convert 0.17 to a fraction.

0.17

Step 1
The entire number becomes the numerator.

$$\frac{17}{100} = \boxed{\frac{17}{100}}$$

Step 2
The position, or place value of the number farthest to the right indicates the denominator. (The 7 is in the <u>hundredths</u> place.)

Step 3—
Reduce the fraction, if possible.

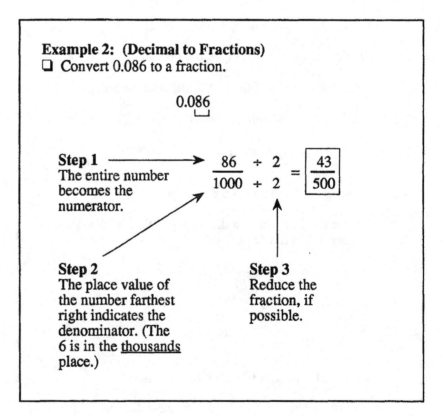

Example 2: (Decimal to Fractions)
❑ Convert 0.086 to a fraction.

0.086

Step 1
The entire number becomes the numerator.

$$\frac{86}{1000} \begin{array}{c} \div 2 \\ \div 2 \end{array} = \boxed{\frac{43}{500}}$$

Step 2
The place value of the number farthest right indicates the denominator. (The 6 is in the <u>thousands</u> place.)

Step 3
Reduce the fraction, if possible.

---

*Refer to Section 4.1 for a review of place values.
**Refer to Chapter 3, Section 3.3 for a discussion of reducing fractions

Example 3: (Decimals to Fractions)
❏ Convert 27.9 to an improper fraction.

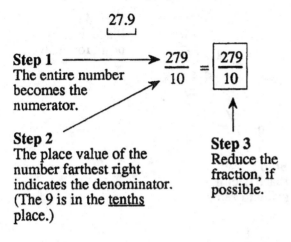

27.9

Step 1 ⟶ $\dfrac{279}{10}$ = $\boxed{\dfrac{279}{10}}$
The entire number
becomes the
numerator.

Step 2
The place value of the
number farthest right
indicates the denominator.
(The 9 is in the <u>tenths</u>
place.)

Step 3
Reduce the
fraction, if
possible.

CONVERTING DECIMALS TO MIXED NUMBERS

Decimal numbers greater than one may be converted to fractions, as shown in Example 3, or to mixed numbers, shown in Examples 4 and 5.

To convert a decimal to a mixed number:

1. Write the whole number (the number to the left of the decimal point).

2. The number to the right of the decimal point is the numerator of the fraction.

3. The position (place value) of the last number to the right of the decimal point indicates the denominator of the fraction.

4. Reduce the fraction, if possible.

Example 4: (Decimals to Mixed Numbers)
❏ Convert 192.75 to a mixed number.

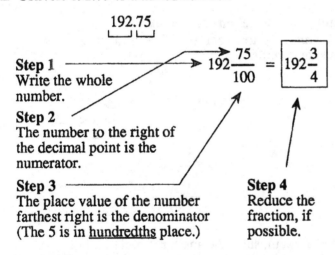

192.75

Step 1 ⟶ $192\dfrac{75}{100}$ = $\boxed{192\dfrac{3}{4}}$
Write the whole
number.

Step 2
The number to the right of
the decimal point is the
numerator.

Step 3
The place value of the number
farthest right is the denominator
(The 5 is in <u>hundredths</u> place.)

Step 4
Reduce the
fraction, if
possible.

Example 5: (Decimals to Mixed Numbers)
❏ Convert 7.005 to a mixed number.

7.005 = $7\dfrac{5}{1000}$ = $\boxed{7\dfrac{1}{200}}$

Whole
Number   Fraction

---

*An improper fraction is a fraction which has a numerator greater than its denominator, such as 27/10.
Mixed numbers are numbers comprised of a whole number and a fraction, such as 2-7/10. Refer to Chapter 3, Section 3.5.

CONVERTING FRACTIONS TO DECIMALS

When converting a fraction to a decimal:

1. Put the fraction in division form (the denominator is the divisor and the numerator is placed under the division sign).

2. Place a decimal point at the end of the numerator and add two or three zeros, if needed. (The number of zeros depends on whether you wish to carry out your answer to tenths hundredths, etc.)

3. Place the decimal point for your answer directly above the numerator decimal point.

4. Complete the division, rounding as needed.

Always complete your division one place farther to the right than you wish the final answer. This will allow you to round your answer to the desired place value.

Example 1: (Fractions to Decimals)
❏ Convert 1/8 to a decimal.

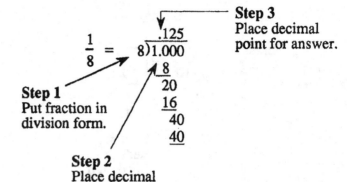

Step 1
Put fraction in division form.

Step 2
Place decimal and add zeros.

Step 3
Place decimal point for answer.

$$\frac{1}{8} = \boxed{0.125}$$

Example 2: (Fractions to Decimals)
❏ Convert 7/16 to a decimal. Give the answer to the nearest hundredth.

$$\frac{7}{16} = \begin{array}{r} .437 \\ 16\overline{)7.000} \\ \underline{64} \\ 60 \\ \underline{48} \\ 120 \\ \underline{112} \\ 80 \end{array} = \begin{array}{l} 0.44 \\ \text{Rounded to} \\ \text{hundredths} \end{array}$$

In this problem, since the answer is requested in hundredths, continue your division to the next place to the right than you wish the final answer. This will allow you to round your answer to the desired place value.

Example 3: (Fractions to Decimals)
❑ Convert 124/17 to a decimal. Round the answer to the nearest tenth.

$$\frac{124}{17} \quad = \quad 17\overline{)\begin{matrix}7.29 \\ 124.00\end{matrix}} \quad = \quad \boxed{\begin{matrix}7.3 \\ \text{Rounded to} \\ \text{nearest tenth}\end{matrix}}$$

$$\begin{matrix}\underline{119} \\ 50 \\ \underline{34} \\ 160 \\ \underline{153} \\ 7\end{matrix}$$

Example 4: (Mixed Numbers to Decimals)
❑ Convert 9-1/6 to a decimal. Give the answer to the nearest hundredth.

$$9\frac{1}{6} = \boxed{9.17} \longleftarrow \textbf{Step 2}$$

Convert the fraction to a decimal:

Step 1
Write the whole number followed by a decimal point.

$$\frac{1}{6} = 6\overline{)\begin{matrix}.166 \\ 1.000\end{matrix}} = .17$$

$$\begin{matrix}\underline{6} \\ 40 \\ \underline{36} \\ 4\end{matrix}$$

MIXED NUMBERS TO DECIMALS

Converting a mixed number to a decimal is similar to converting a fraction to a decimal. The only difference is that the whole number (the number to the left of the decimal point) is already known. To convert a mixed number to a fraction:

1. Write the whole number given in the mixed number and place a decimal point at the end of it.

2. Convert the fraction to a decimal (as shown in Examples 1-3) and record the numbers to the right of the decimal point.

Example 5: (Mixed Numbers to Decimals)
❑ Convert 8-5/7 to a decimal. Round to the nearest tenth.

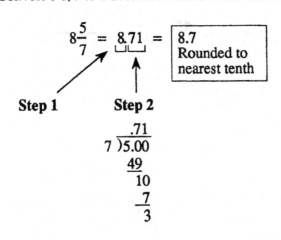

$$8\frac{5}{7} = 8.71 = \boxed{\begin{matrix}8.7 \\ \text{Rounded to} \\ \text{nearest tenth}\end{matrix}}$$

Step 1   **Step 2**

$$7\overline{)\begin{matrix}.71 \\ 5.00\end{matrix}}$$

$$\begin{matrix}\underline{49} \\ 10 \\ \underline{7} \\ 3\end{matrix}$$

PRACTICE PROBLEMS 4.5: Converting Decimals to Fractions

❏ Convert the following decimals to fractions. (Do not reduce the fractions.)

1. 0.125 = _____   **2.** 0.78 = _____

❏ Convert the following decimals to mixed numbers. (Do not reduce the fractions.)

3. 1.75 = _____   **4.** 53.625 = _____   **5.** 195.56 = _____

❏ Convert the fractions to decimals.

6. $\dfrac{4}{15}$ = _____   **7.** $\dfrac{2}{3}$ = _____

❏ Convert the mixed numbers shown below to decimals.

8. $25\dfrac{3}{4}$ = _____   **9.** $190\dfrac{1}{8}$ = _____   **10.** $2\dfrac{9}{100}$ = _____

5 *Percents*

SKILLS CHECK

Complete and score the following skills test. Each section should be scored separately in the box provided to the right. A score of 8 or above indicates you are sufficiently strong in that concept. A score of 7 or below indicates a review of that section is advisable.

5.1 Percents and Fractions

Number Correct

❑ Express the following percents as fractions or fractions as percents. Be sure fractions are given in lowest terms.

1. 40% = _____

2. 62% = _____

3. $\dfrac{2}{10}$ = _____

4. 34% = _____

5. $\dfrac{6}{20}$ = _____

6. $\dfrac{4}{5}$ = _____

7. 17% = _____

8. $\dfrac{6}{5}$ = _____

9. 56% = _____

10. $\dfrac{5}{12}$ = _____

5.2 Percents and Decimals

Number Correct

❑ Express the following percents as decimals or decimals as percents.

1. 19% = _____

2. 0.27 = _____

3. 0.168 = _____

4. 48% = _____

5. 184% = _____

6. 0.729 = _____

7. 0.1536 = _____

8. 66% = _____

9. 0.0529 = _____

10. 97% = _____

5.3 Calculating Percent Problems

❏ Complete the percent problems given below.

1. 220 is what percent of 250? _____

2. What is 6% of 13,530? _____

3. 6 is what percent of 20? _____

4. 58 is 7% of what number? _____

5. What percent is 17,210 of 149,600? _____

6. What number is 5.5% of 172? _____

7. 18% of what number is 29? _____

8. 72% of 2490 is how much? _____

9. 341 is what percent of 8200? _____

10. 4175 is 23% of what number? _____

5.1 PERCENTS AND FRACTIONS

SUMMARY

There are three methods that may be used to convert from a fraction to a percent:

1. Multiply the numerator and denominator by the same number so that the denominator is 100.

$$\frac{6}{25} \times \frac{4}{4} = \frac{24}{100} = \boxed{24\%}$$

2. Set up a proportion and solve for x.

$$\frac{6}{25} = \frac{x}{100}$$

$$\frac{600}{25} = x$$

$$\boxed{24\%} = x$$

3. Divide the fraction to obtain a decimal number. Then convert the decimal number to a percent. (This method is described in Section 5.2.)

The use of percents is a method of comparing one quantity with another. The word **percent** comes from the Latin words *per centum*, meaning "by the hundreds". Mathematically, percent indicates "how many per hundred" or "hundredths". For example, if 4 percent (4%) of the treatment plant staff are ill on a typical day, this means that 4 out of every hundred are ill on a typical day.

This same information can therefore be written mathematically as a percent, a common fraction or a decimal fraction, as shown below:

Percent: 4%

Common Fraction: $\dfrac{4}{100}$

Decimal Fraction: 0.04

There are three methods to convert from a fraction to a percent. Two of the methods involve fractions, while the third involves decimal numbers. This section describes Method 1 and 2. Method 3 is described in Section 5.2.

When converting **from a fraction to a percent**, you may multiply the numerator and denominator by the same number (Method 1 as illustrated in Example 1) or set up and solve a proportion (Method 2, as illustrated in Example 2). The most direct method of converting a fraction to a percent is Method 1, and it should be used whenever possible. When it is not possible, the proportion or decimal method may be used.

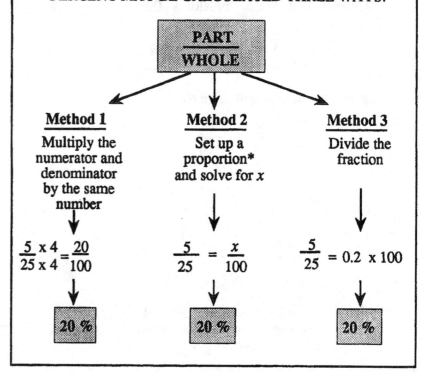

PERCENT IS A CALCULATION OF "HOW MANY PER HUNDRED?"

$$\text{PERCENT} = \frac{?}{100}$$

BEGINNING WITH ANY FRACTION (PART/WHOLE), PERCENT MAY BE CALCULATED THREE WAYS:

$$\frac{\text{PART}}{\text{WHOLE}}$$

Method 1
Multiply the numerator and denominator by the same number

$$\frac{5 \times 4}{25 \times 4} = \frac{20}{100}$$

20 %

Method 2
Set up a proportion* and solve for *x*

$$\frac{5}{25} = \frac{x}{100}$$

20 %

Method 3
Divide the fraction

$$\frac{5}{25} = 0.2 \times 100$$

20 %

Example 1: (Percents and Fractions)
❑ Two out of every twenty District staff live outside the district boundaries. What percent of the staff live outside the district boundaries?

The part of the staff living outside the district compared with the entire district staff (part over whole) is:

$$\frac{2}{20}$$

Since the denominator must be 100 for a percent calculation, multiply the numerator and denominator by 5:

$$\frac{2}{20} \times \frac{\times 5}{\times 5} = \frac{10}{100}$$

$$= \boxed{10\%}$$

---

* To review proportion calculations, refer to Chapter 7.

Example 2: (Percents and Fractions)
❏ Given the fraction 4/17, what percent is this?

Percent is "per 100"; therefore, to determine how many per hundred the fraction 4/17 represents, use a proportion to solve for percent:

$$\frac{4}{17} = \frac{x}{100}$$

$$17x = 400$$

$$x = 24$$

Therefore:

$$\frac{4}{17} = \frac{24}{100}$$

$$\text{or} = \boxed{24\%}$$

Example 3: (Percents and Fractions)
❏ What fraction is equivalent to 32%? (Reduce fraction to lowest terms.)

$$32\% = \frac{32}{100}$$

Once you have the fraction shown above, simply reduce the fraction to lowest terms:

$$\frac{32 \div 4}{100 \div 4} = \boxed{\frac{8}{25}}$$

When converting **from a percent to a fraction,** you simply write the percent as a common fraction ($x\% = x/100$), then reduce the fraction*. For example:

$$28\% = \frac{28}{100} = \frac{7}{25}$$

Example 3 illustrates this calculation.

---

* Refer to Chapter 3 for a review of reducing fractions.

PRACTICE PROBLEMS 5.1: Percents and Fractions

❏ Convert the following fractions to percents, or vice versa, as indicated.

1. $\dfrac{15}{25}$ = _____

6. 6% = _____

2. 16% = _____

7. $\dfrac{4}{20}$ = _____

3. 48% = _____

8. 3% = _____

4. $\dfrac{6}{15}$ = _____

9. $\dfrac{1}{7}$ = _____

5. $\dfrac{3}{75}$ = _____

10. 74% = _____

5.2 PERCENTS AND DECIMALS

SUMMARY

To convert from a decimal to a percent:

1. Write the decimal fraction as a common fraction with 100 as the denominator.

$$0.46 = \frac{46}{100}$$

2. Then write the fraction as a percent.

$$\frac{46}{100} = \boxed{46\%}$$

3. These two steps are often combined into one step. **Simply move the decimal point two places to the right:**

$$0.46 = \boxed{46\%}$$

To convert from a percent to a decimal:

1. Write the percent as a common fraction, with 100 as the denominator.

$$82\% = \frac{82}{100}$$

2. Then write the common fraction as a decimal fraction.

$$\frac{82}{100} = 0.82$$

3. These two steps are often combined into one step. **Simply move the decimal point two places to the left:**

$$82.\% = \boxed{0.82}$$

CONVERTING A DECIMAL NUMBER TO A PERCENT

When converting from a decimal to a percent:

- First, write the decimal number as a common fraction with one hundred as the denominator;

- Then write the fraction as a percent.

Often these two steps are combined into one. Since percent refers to "hundredths," if you know how many hundredths, you know the percent. For example 0.37 is read as "32 hundredths". It is therefore equivalent to "32 percent";

$$0.32 = \frac{32}{100} = 32\%$$

Note the position of the decimal point in the decimal and percent numbers. The decimal point in the percent number is always **two places further to the right** than the same number expressed as a decimal. For this reason, you will often see the conversion from a decimal to a percent written as:

$$\boxed{\frac{\text{Decimal}}{\text{Number}} \times 100 = \%}$$

Multiplication by 100 results in a decimal point move **two places to the right**.

Example 1: (Percents and Decimals)
❑ What is 0.71 expressed as a percent?

One method to determine percent is to express the decimal as a fraction with a denominator of 100:

$$0.71 = \frac{71}{100}$$

Then write the fraction as a percent:

$$\frac{71}{100} = \boxed{71\%}$$

The most direct method of solving this problem is to use the fact that "hundredths" means "percent". Therefore 71 hundredths is 71%:

$$0.71 = \frac{71}{100} = \boxed{71\%}$$

Example 2: (Percents and Decimals)
❑ What is 1.05 expressed as a percent?

The direct method will be used to determine percent in this problem. Since 1.05 is read as 105 hundredths, then it must be equal to 105 percent:

$$1.05 = \frac{105}{100} = \boxed{105\%}$$

Example 3: (Percents and Decimals)
❑ Given the fraction 7/15, express this fraction as a percent.

Since the denominator of the fraction cannot be multiplied by a whole number to equal 100, divide the numerator by the denominator to obtain a decimal number:

$$
\begin{array}{r}
0.466 \\
15\overline{)7.000} \\
\underline{60} \\
100 \\
\underline{90} \\
100 \\
\underline{90}
\end{array}
$$

This decimal number can now be converted to a percent. As shown below, 47 hundredths is equivalent to 47 percent:

$$0.47 = \frac{47}{100} = \boxed{47\%}$$

Example 4: (Percents and Decimals)
❑ Convert 2% to a decimal number.

The word "percent" means hundredths. Therefore, 2% may be written as a decimal number:

$$2\% = \frac{2}{100} = \boxed{0.02}$$

CONVERTING A FRACTION TO A DECIMAL NUMBER TO A PERCENT

As discussed in the previous section, the most direct method to convert from a fraction to a percent is to multiply the numerator and denominator by a whole number so that the denominator will equal 100. However, many times this is not possible. In this situation, the most common way to determine percent is to divide the fraction and then convert the resulting decimal number to a percent. For example:

$$\frac{4}{17} = 0.235 = \frac{23.5}{100} = 23.5\%$$

CONVERTING A PERCENT BACK TO A DECIMAL NUMBER

To convert from a percent to a decimal number, it is helpful to remember that **percent** refers to **hundredths**. Therefore, if you know the percent, you also know the number as a decimal to the hundredths place*. For example, 16 percent is equivalent to 16 hundredths. This is written mathematically as:

$$16\% = \frac{16}{100} = 0.16$$

Again, note the position of the decimal point in the percent and decimal numbers. Division by 100 results in a decimal point move **two places to the left**.

---

* For a review of the place value system, refer to Chapter 4, Decimals.

PRACTICE PROBLEMS 5.2: Percents and Decimals

❑ Express the following decimals as percents.

1. 0.643 = _____   **4.** 0.753 = _____

2. 0.0346 = _____   **5.** 0.057 = _____

3. 1.02 = _____

❑ Express the following percents as decimals.

6. 19% = _____   **9.** 4% = _____

7. 87% = _____   **10.** 0.5% = _____

8. 108% = _____

5.3 CALCULATING PERCENT PROBLEMS

SUMMARY

The following equation may be used to calculate percent:

$$\% \;=\; \frac{\text{Part}}{\text{Whole}} \times 100$$

Several types of water and wastewater calculations require an understanding of percent calculations. Many of these applied calculations are discussed in Chapter 6 of the applied math texts—Efficiency and other Percent Calculations. The basic concept of the percent problem is presented here.

Percent is a useful tool for comparing data or expressing data in terms more easily understood. For example, to say that suspended solids removal is 120 mg/*L* really doesn't say much. Was the 120 mg/*L* removed from an original content of 340 mg/*L* or from an original content of 200 mg/*L*? Obviously, a removal of 120 mg/*L* suspended solids from a 200 mg/*L* content is a much better removal rate than from a 340 mg/*L* suspended solids content. The percent calculation makes this comparison easier:

$$\% = \frac{Part}{Whole} \times 100$$

$$\% = \frac{120 \text{ mg/}L}{200 \text{ mg/}L} \times 100$$

$$= \boxed{60\% \text{ Removal}}$$

$$\% = \frac{120 \text{ mg/}L}{340 \text{ mg/}L} \times 100$$

$$= \boxed{35\% \text{ Removal}}$$

THE SAME EQUATION CAN BE USED TO CALCULATE MANY TYPES OF PERCENT PROBLEMS

$$\% = \frac{Part}{Whole} \times 100$$

Example 1: (Calculating Percent Problems)
❑ 17 is what percent of 54? (Round to the nearest whole percent.)

The most important aspect in using the percent equation is knowing which number represents the "part" and which represents the "whole". It is helpful to know that in most cases, the number that follows "of" is the "whole:"

$$\% = \frac{Part}{Whole} \times 100$$

$$x = \frac{17}{54} \times 100$$

$$x = \boxed{31\%}$$

Example 1: (Calculating Percent Problems)
❑ What is 5% of 1840?

The unknown value* is this problem is the "part". The "whole" is 1840:

$$\% = \frac{Part}{Whole} \times 100$$

$$5 = \frac{x}{1840} \times 100$$

$$\frac{(5)(1840)}{100} = x$$

$$\boxed{92} = x$$

---

* For a review of Solving for the Unknown Value, refer to Chapter 2.

Example 3: (Calculating Percent Problems)
❑ What is 78% of 435? (Round to the nearest whole number.)

$$\% = \frac{\text{Part}}{\text{Whole}} \times 100$$

$$78 = \frac{x}{435} \times 100$$

$$\frac{(78)(435)}{100} = x$$

$$\boxed{340} = x$$

Example 4: (Calculating Percent Problems)
❑ 125 is what percent of 642? (Round to the nearest whole percent.)

$$\% = \frac{\text{Part}}{\text{Whole}} \times 100$$

$$x = \frac{125}{642} \times 100$$

$$x = \boxed{19\%}$$

PERCENT VS. PERCENTAGE

Often the terms percent and percentage are confused and therefore when the percent equation is used, numbers may be put in incorrect positions in the equation.

Percent, sometimes called the rate, is placed in the % position in the equation.

Percentage on the other hand, is the "part" in the equation,

$$\% = \frac{\text{Part}}{\text{Whole}} \times 100$$

PRACTICE PROBLEMS 5.3: Calculating Percent Problems

❑ Calculate the unknown value in each of the problems given below.

1. 68% of 2140 is how much? _____

2. What number is 6.3% of 213? _____

3. 167 is 4% of what number? _____

4. 4900 is 19% of what number? _____

5. 292 is what percent of 7810? _____

6. 28% of what number is 53? _____

7. What percent is 12,940 of 96,520? _____

8. 9 is what percent of 48? _____

9. What is 3% of 8720? _____

10. 219 is what percent of 302? _____

6 *Averages*

Complete and score the following skills test. Each section should be scored separately in the box provided to the right. A score of 4 or above indicates you are sufficiently strong in that concept. A score of 3 or below indicates a review of that section is advisable.

6.1 Averages—Arithmetic Mean

❑ Calculate the average (mean) value in each problem below.

1.

mg/L	mg/L
140	131
136	148
152	159
	166

= _____

2.

17 mg/L	19 mg/L
15 mg/L	10 mg/L

= _____

3.

MGD	MGD
1.32	1.52
1.67	1.05
1.48	1.21
	1.38

= _____

4.

lbs	lbs
6340	4863
5492	5124
5976	5618
	4766

= _____

5.

MGD	MGD
0.061	0.048
0.052	0.056
0.068	0.063
	0.069

= _____

Number Correct ☐

Averages—Median

❑ Determine the median value in each problem given below.

1.

MPN/100 ml	
220	3520
230	280
230	250
220	220
310	240
290	

= _____

2.

mg/L	mg/L
162	156
150	162
168	183
175	178

= _____

3.

SVI	SVI
117	125
117	129
128	115
123	

= _____

Number Correct ☐

(Continued on next page.)

Averages—Median (Cont'd)

4. MPN/100 ml: 230, 250, 230, 220, 270, 840, 290

= _____

5. mg/*L*: 114, 128, 121, 136, 122, 115, 126, 133, 124, 145, 133

= _____

Averages—Mode

❏ Find the mode value for each of the problems given in the Median Section.

1. _____

2. _____

3. _____

4. _____

5. _____

6.2 Moving Averages

❏ Calculate a two-day moving average, beginning with the first two values in problems 1-3.

1.

mg/*L*	Moving Average
20	
17	_____
21	_____
24	_____
18	_____
19	_____

2.

mg/*L*	Moving Average
315	
295	_____
305	_____
292	_____
286	_____
291	_____

6.2 Moving Averages—Cont'd

3.

mg/L	Moving Average
28	
21	_____
16	_____
24	_____

❏ Calculate a 7-day moving average beginning with the first seven values in each problem below.

4.

mg/L	Moving Average
2460	
2190	
2640	
2210	
2540	
2820	
2750	_____
2680	_____
2490	_____
2820	_____

5.

mg/L	Moving Average
216	
228	
220	
219	
231	
228	
219	_____
214	_____
221	_____
226	_____

6.3 Weighted Averages

❏ Organize the data below in five-point spans and calculate the weighted average.

1. Begin with the span 206-210.

mg/L	mg/L	mg/L
218	223	209
239	216	211
232	213	224
221	226	208
242	234	216
236	236	231
214	240	228
223	227	220
219	221	233
230	217	225

= _____

2. Begin with the span 116-120.

mg/L	mg/L	mg/L
150	122	136
128	127	133
136	139	121
141	146	128
131	151	117
126	144	120
129	142	129
147	137	135
153	150	130
130	141	122

= _____

Number
Correct

6.3 Weighted Averages—Cont'd

❑ Calculate a weighted average using a ten-point span in grouping data.

3. Begin with the span 2391-2400.

mg/L	mg/L	mg/L
2470	2414	2530
2519	2505	2515
2475	2482	2472
2492	2470	2485
2515	2520	2426
2465	2483	2415
2398	2450	2397
2449	2426	2405
2460	2434	2419
2528	2410	2427

= _____

4. Begin with the span 111-120.

mg/L	mg/L	mg/L
127	132	136
121	124	129
114	128	121
125	121	124
132	119	132
137	115	139
128	123	145
131	129	151
139	133	148
142	138	132

= _____

5. Begin with the span 11-20.

mg/L	mg/L	mg/L
21	19	17
18	22	16
15	28	20
19	22	21
23	29	15
29	30	19
32	31	26
30	27	29
25	21	36
18	19	42

= _____

6.1 Averages

```
┌─────────────────────────────────────────────────┐
│                    SUMMARY                        │
├─────────────────────────────────────────────────┤
```

1. To calculate an **arithmetic mean**:

$$\text{Mean} = \frac{\text{Sum of All Measurements}}{\text{Number of Measurements Used}}$$

2. To determine the **median** value of a set of numbers:

   - Arrange the numbers according to size.

   - The middle value is the median value.

3. The **mode** value is the value that occurs most frequently.

In order to evaluate the day-to-day or overall performance of a treatment process, much data must be collected and evaluated. Because there may be much variation in the information, it is often difficult to determine trends in performance.

By calculating an average, a group of data is represented by a single number. This number may be considered typical of the group. Three expressions of averages are the mean (sometimes called "arithmetic mean"), the median, and the mode.

The arithmetic mean is the most commonly used measurement of average value, and most calculations of average are those of arithmetic mean rather than the median or mode average.

When evaluating information based on averages, remember that the "average" reflects the general nature of the group and does not necessarily reflect any one element of that group.

The arithmetic mean is calculated as follows:

$$\frac{\text{Arithmetic}}{\text{Mean}} = \frac{\text{Sum of All Meas.}}{\text{No. of Meas. Used}}$$

When the word "average" is used in a math problem, it refers to the **arithmetic mean,** unless otherwise stated.

Example 1: (Averages)

❏ For the primary influent flow, the following composite-sampled suspended solids concentrations were recorded for the week: Monday—320 mg/L; Tuesday—317 mg/L; Wednesday—308 mg/L; Thursday—313 mg/L; Friday—325 mg/L; Saturday—316 mg/L; Sunday—310 mg/L. Calculate the average daily suspended solids concentration.

1.	Monday	320 mg/L SS
2.	Tuesday	317 mg/L SS
3.	Wednesday	308 mg/L SS
4.	Thursday	313 mg/L SS
5.	Friday	325 mg/L SS
6.	Saturday	316 mg/L SS
7.	Sunday	310 mg/L SS
	Total	2209 mg/L SS

$$\frac{\text{Average SS}}{\text{Concentration}} = \frac{\text{Sum of All Measurements}}{\text{Number of Measurements Used}}$$

$$= \frac{2209 \text{ mg/L SS}}{7}$$

$$= \boxed{316 \text{ mg/L} \atop \text{SS}}$$

Example 2: (Averages)

❏ BOD test results from daily composite samples taken on a trickling filter influent are as follows: 210 mg/L, 190 mg/L, 230 mg/L, 250 mg/L, 235 mg/L, and 220 mg/L. What is the average BOD concentration for this 6-day period?

	210 mg/L
	190 mg/L
	230 mg/L
	250 mg/L
	235 mg/L
	220 mg/L
Total	1335 mg/L

$$\frac{\text{Average BOD}}{\text{Concentration}} = \frac{\text{Sum of All Measurements}}{\text{Number of Measurements Used}}$$

$$= \frac{1335 \text{ mg/L}}{6}$$

$$= \boxed{223 \text{ mg/L} \atop \text{BOD}}$$

Example 3: (Averages)
❏ The number of sick days taken during the year by each of 5 operators was recorded as follows: Operator A—6 days; Operator B—3 days: Operator C—7 days; Operator D—0 days and Operator E—1 day. What was the average number of days sick leave taken by these five operators?

1.	Operator A—	6	days sick leave
2.	Operator B—	3	days sick leave
3.	Operator C—	7	days sick leave
4.	Operator D—	0	days sick leave
5.	Operator E—	1	days sick leave
	Total	17	days sick leave

$$\text{Average Days Sick Leave} = \frac{\text{Sum of All Measurements}}{\text{Number of Measurements Used}}$$

$$= \frac{17 \text{ days}}{5}$$

$$= \boxed{3.4 \text{ days sick leave}}$$

AVERAGE FLOW USING TOTALIZER READINGS

The average flow for any time period (hours, days, etc.) may be determined using totalizer readings at the beginning and end of the desired time period. For example, to calculate average hourly flow:

$$\text{Aver. Hourly Flow (gph)} = \frac{\text{Tot. Flow during Period}}{\text{Number of Hours}}$$

And average daily flow can be calculated as:

$$\text{Aver. Daily Flow (gpd)} = \frac{\text{Tot. Flow during Period}}{\text{Number of Days}}$$

Example 4: (Averages)
❏ The plant totalizer reading is 81,426,172 gallons on June 1st. On June 30th, the totalizer reading is 106,720,712 gallons. What is the average daily flow (in MGD) for the month of June?

$$\text{Average Daily Flow} = \frac{\text{Sum of All Daily Flows}}{\text{Number of Daily Flows}}$$

In this problem, the total of all daily flows can be determined by subtracting the second totalizer reading from the first totalizer reading (106,720,712 − 81,426,172 = 25,294,540 gallons):

$$\text{Average Daily Flow} = \frac{25,294,540 \text{ gal}}{30 \text{ days}}$$

$$= \frac{843,151 \text{ gal}}{\text{day}}$$

$$= \boxed{0.843 \text{ MGD}}$$

DETERMINING THE MEDIAN VALUE

A median is another type of average and, in contrast with the arithmetic mean, there is little or no mathematical calculation associated with it. The median average is the **middle value of all measurements taken**. Because 50% of the values will be below the median, we must **first arrange the sample values in ascending or descending order**. Once this is done, finding the median (middle value) is not difficult.

Example 5: (Averages)

❑ At a treatment plant, the following cu ft/day screening withdrawals were recorded for the week: Monday—3.8 cu ft; Tuesday—3.5 cu ft; Wednesday—3.3 cu ft; Thursday—3.6 cu ft; Friday—3.5 cu ft; Saturday—7.1 cu ft; Sunday—3.9 cu ft. What is the median daily screening withdrawal at the plant?

To determine the median value, first arrange the numbers in ascending or descending order. Then the middle value can be determined easily:

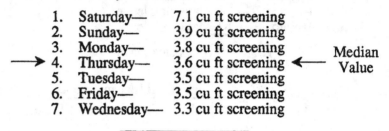

1.	Saturday—	7.1 cu ft screening	
2.	Sunday—	3.9 cu ft screening	
3.	Monday—	3.8 cu ft screening	Median
→ 4.	Thursday—	3.6 cu ft screening ←	Value
5.	Tuesday—	3.5 cu ft screening	
6.	Friday—	3.5 cu ft screening	
7.	Wednesday—	3.3 cu ft screening	

> 3.6 cu ft screening
> median average

Example 6: (Averages)

❑ The MGD flow rates for a 13-day period are as follows: 2.67, 3.61, 4.05, 2.32, 2.08, 1.97, 2.17, 2.25, 2.81, 3.02, 3.92, 3.21, and 3.18. What is the median flow rate for this two-week period?

First, arrange the values in descending order, then find the middle value:

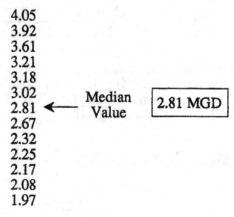

4.05
3.92
3.61
3.21
3.18
3.02
2.81 ← Median Value   2.81 MGD
2.67
2.32
2.25
2.17
2.08
1.97

Example 7: (Averages)

❏ The SVI (sludge volume index) values during a six-day period for an activated sludge plant were 118, 122, 116, 126, 111, 129. What is the median SVI value for the six-day period?

First arrange the values in descending order:

```
                                   129
                                   126
The median is          ────→       122 ⎤  Middle
halfway between                    118 ⎦  numbers
these numbers                      116
                                   111
```

Example 8: (Averages)

❏ A wastewater sample is tested for coliform bacteria using the multiple-tube coliform bacteria test. Given the results shown below, calculate the mean and median values. Which value best reflects typical values?

*MPN/100 ml: 210, 240, 270, 230, 260, 1540, 220

First calculate the mean value:

$$\frac{\text{Mean Value}}{\text{MPN/100 ml}} = \frac{2970}{7}$$

$$= \boxed{424}$$

Then determine the median value by arranging the data in descending order:

```
        1540
         270
         260
         240  ◄──── Median Value
         230
         220        ┌─────┐
         210        │ 240 │
                    └─────┘
```

In this example, the median value more accurately reflects the typical value.

FINDING THE MEDIAN VALUE BETWEEN TWO NUMBERS

Whenever there is an even number of measurements, there are two numbers in the middle position, as shown in Example 7. The median value is the value **halfway between the two middle numbers**.

There are two methods that may be used to calculate the value halfway between two numbers:

1. Add the two numbers and divide by 2. For example,

$$\frac{122 + 118}{2} = \frac{240}{2} = \boxed{120}$$

2. Find the difference between the two numbers (using subtraction), divide that distance by 2, then add the result to the smaller number. Using the same example:

$$122 - 118 = 4 \text{ (Difference)}$$

$$4 \div 2 = 2 \text{ (Half the Difference)}$$

$$118 + 2 = \boxed{120}$$

COMPARING MEAN AND MEDIAN VALUES

An "average value" is supposed to represent a value **typical** of a group of values. The arithmetic mean is most often the preferred expression of average. However, when there are extreme values (unusually low or unusually high), the median value is more representative of the typical value. This is because the calculation of arithmetic mean must include the extreme value, and the answer is therefore pulled in the direction of that extreme value. Example 8 illustrates the comparison of mean and median values when an extreme value is included.

---

* Most probable number of coliform group bacteria.

DETERMINING THE MODE VALUE

The mode is the value that occurs most frequently. If there are no two identical values, then no mode exists. There may be one mode, two modes (bimodal) or more, as shown in Examples 9-12.

Example 9: (Averages)
❑ Given the data below, determine the mode value(s).

210 mg/L
190 mg/L
230 mg/L
250 mg/L
230 mg/L
220 mg/L

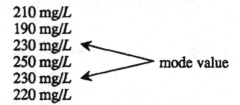 mode value

The mode value is 230 mg/L since it occurs twice. There is no other mode in this set of numbers.

Example 10: (Averages)
❑ What is the mode value(s) for the following data:

mg/L	mg/L
220	270
210	232
235	219
242	220
220	224
251	240
243	232

The mode for this data is:

220 mg/L (occurs three times)

Example 11: (Averages)
❑ Determine the mode value(s) for the following data:
165, 180, 120, 155, 140, 180, 195, 170, 150, 110, 130, 165, 125, 160.

There are two modes for this data:

165 (occurs twice)
180 (occurs twice)

Example 12: (Averages)
❑ Determine the mean, median, and mode values of the following coliform bacteria test results.

*MPN/100 ml: 230, 210, 230, 290, 340, 2450, 310

First calculate the arithmetic mean:

$$\frac{\text{Mean Value}}{\text{MPN/100 ml}} = \frac{\text{Sum of All Measurements}}{\text{Number of Measurements}}$$

$$= \frac{4060}{7}$$

$$= \boxed{580}$$

Next determine the median value by arranging the data in descending order:

2450, 340, 310, $\boxed{290,}$ 230, 230, 210

↑

Median Value

Then determine the mode value:

$\boxed{230}$ is the mode value

---

* MPN refers to the most Probable Number of coliform group bacteria.

PRACTICE PROBLEMS 6.1: Averages

❑ Calculate the arithmetic mean for each problem below.

1.

mg/L	mg/L
146	149
130	158
151	164
132	

= _____

2.

MGD
0.91
0.83
0.97
1.36
1.15
0.84

= _____

3.

mg/L
2470
2185
2316

= _____

4.

lbs	lbs
4620	4780
4130	4640
4910	4270
5300	4480

= _____

5.

mg/L	mg/L
42	45
48	41
39	49
36	52

= _____

❑ Determine the arithmetic mean and median for problems 6-8.

6.

MPN/100
260
220
240
290
360
3310
415
280
240

= _____

7.

mg/L
170
190
180
240
190
160
175

= _____

8.

mg/L
2450
2610
2290
2540
2650
1820
2210

= _____

❑ Determine the mode value for the following data sets:

9. Mode value for problem 6

= _____

10. Mode value for problem 7

= _____

6.2 MOVING AVERAGES

<table>
<tr><td colspan="2" align="center">**SUMMARY**</td></tr>
<tr><td>

To calculate a moving average, use the same basic arithmetic mean equation:

$$\text{Arithmetic Mean} = \frac{\text{Sum of All Measurements}}{\text{Number of Measurements Used}}$$

The only difference in these calculations is that the "data window" moves from day to day.

</td></tr>
</table>

A moving average is a calculation of arithmetic mean. The most common type of moving average is the 7-day moving average. Each day the 7-day average drops the oldest value and adds the newest value. Therefore, the "7-day window" of data shifts each day.

A "moving average" is simply a special type of arithmetic mean calculation. Therefore, the same equation is used as when calculating any other arithmetic mean:

$$\text{Arithmetic Mean} = \frac{\text{Sum of All Meas.}}{\text{No. of Meas. Used}}$$

For moving average calculations, there is a set time frame such as 7 days or 2 days. Each day, the oldest value is dropped, the newest value is added, and a new arithmetic mean is then calculated. Examples 1 and 2 illustrate this calculation.

The 7-day moving averages indicate trends in system operation since they tend to "smooth out" the more pronounced day-to-day fluctuations in data.

THE "DATA WINDOW" SHIFTS EVERY DAY FOR A MOVING AVERAGE

Day	mg/L
1	220
2	245
3	236
4	219
5	216
6	231
7	228
8	245
9	251
10	248

1st 7-day Average
2nd 7-day Average
3rd 7-day Average

Example 1: (Moving Averages)
❑ Given the data below, calculate the first two 7-day moving averages.

Day	mg/L	Day	mg/L	Day	mg/L
1	170	5	184	9	174
2	182	6	172	10	192
3	164	7	186	11	186
4	178	8	167	12	179

The first 7-day moving average includes the first 7 days of data:

$$\text{Arithmetic Mean} = \frac{\text{Sum of All Measurements}}{\text{Number of Measurements Used}}$$

$$= \frac{1236}{7}$$

$$= \boxed{177 \text{ mg/}L}$$

To obtain the second 7-day average, drop the value from Day 1 (170 mg/L) and add the value from Day 8 (167 mg/L):

$$\text{2nd 7-day Average} = \frac{1236 - 170 + 167}{7}$$

$$= \frac{1233}{7}$$

$$= \boxed{176 \text{ mg/}L}$$

Example 2: (Moving Averages)

❑ Using the data provided below, calculate the first four 7-day moving averages.

Day	mg/L	Day	mg/L	Day	mg/L
1	215	6	210	11	224
2	237	7	219	12	217
3	220	8	229	13	225
4	234	9	221		
5	217	10	236		

Calculate the first 7-day moving average using data for the first 7 days:

$$\text{Arithmetic Mean} = \frac{\text{Sum of All Measurements}}{\text{Number of Measurements Used}}$$

$$= \frac{1552}{7}$$

$$= \boxed{222 \text{ mg/L}}$$

To calculate the second 7-day moving average, drop the data from Day 1 and add the data from Day 8:

$$\text{2nd 7-day Average} = \frac{1552 - 215 + 229}{7}$$

$$= \frac{1566}{7}$$

$$= \boxed{224 \text{ mg/L}}$$

The third and fourth 7-day moving averages are calculated similarly—dropping the oldest value and adding a new value:

$$\text{3rd 7-day Average} = \frac{1566 - 237 + 221}{7}$$

$$= \frac{1550}{7}$$

$$= \boxed{221 \text{ mg/L}}$$

$$\text{4th 7-day Average} = \frac{1550 - 220 + 236}{7}$$

$$= \frac{1566}{7}$$

$$= \boxed{224 \text{ mg/L}}$$

PRACTICE PROBLEMS 6.2: Moving Averages

❑ Calculate the first five 7-day moving averages for each problem given below.

1.

Day	SVI	Day	SVI
1	110	7	133
2	105	8	126
3	113	9	131
4	123	10	122
5	140	11	124
6	117		

= _____ _____

_____ _____

_____

2.

Day	mg/L	Day	mg/L
1	260	8	308
2	225	9	297
3	232	10	269
4	221	11	273
5	246	12	256
6	283	13	281
7	324	14	262

= _____ _____

_____ _____

_____

❑ Calculate the first four 5-day moving averages for each problem given below.

3.

Day	mg/L	Moving Average
1	24	
2	18	
3	16	
4	21	
5	28	_____
6	24	_____
7	33	_____
8	22	_____

4.

Day	mg/L	Moving Average
1	136	
2	152	
3	145	
4	167	
5	151	_____
6	174	_____
7	149	_____
8	136	_____

6.3 WEIGHTED AVERAGES

SUMMARY

To calculate a weighted average:

1. Establish the point span of the desired grouping.

2. List how many data points fall into each grouping. (This is referred to as the "frequency".)

3. Determine the average value for each grouping, then multiply the average value by the frequency of that group.

4. Divide the sum of the products by the frequency to obtain the average value.

$$\text{Weighted Average} = \frac{\text{Sum of the Products}}{\text{Total Frequency}}$$

A weighted average is sometimes used in determining an average value for a large set of data.

When there is a large set of data to be averaged, a weighted average is sometimes used to determine an average value for the data.

The data is first arranged into groups, such as 5-point groupings, 10-point groupings, etc., depending on the point span of the original data.

Although there is some loss in accuracy, as long as the point span is not too great, the weighted average obtained is normally very close to the average that would be obtained using the normal arithmetic mean calculation.

LARGE SETS OF DATA CAN BE GROUPED TO DETERMINE AVERAGE VALUE

Group*	Frequency		Grp. Aver.		Product
115-119	2	x	117	=	234
120-124	4	x	122	=	488
125-129	4	x	127	=	508
130-134	1	x	132	=	132
	11				1362

Step 1

Establish the size groups. These groups have a 5-pt. span.

Step 2

List how many data points fall into each grouping.

Step 3

Determine the average value for each group. (It will be the middle value of the group.)

Step 4

Multiply the group average by the frequency.

Step 5

As with other calculations of arithmetic mean, the sum of all measurements is divided by the number of measurements (frequency):

$$\text{Arithmetic Mean} = \frac{\text{Sum of All Measurements}}{\text{Number of Measurements Used}}$$

Using the column titles shown above, this could be restated as:

$$\text{Weighted Average} = \frac{\text{Sum of the Products}}{\text{Total Frequency}}$$

$$= \frac{1362}{11} = \boxed{124}$$

---

* This is only a sample of a much larger data base. A weighted average calculation would not normally be used to calculate only 11 data points.

Example 1: (Weighted Averages)
❏ Group the data given below in 10-point groups and calculate a weighted average. Then calculate a regular arithmetic mean and compare values.

244	190	192	249	182	250
221	189	185	211	178	247
252	202	174	236	194	214
198	218	227	215	207	201
231	196	248	256	262	186

List the groups and frequency, then calculate the average value:

Group	Freq*		Grp. Aver.		Product
170-179	l l	x	174.5	=	349
180-189	l l l l	x	184.5	=	738
190-199	l l l l l	x	194.5	=	972.5
200-209	l l l	x	204.5	=	613.5
210-219	l l l l	x	214.5	=	858
220-229	l l	x	224.5	=	449
230-239	l l	x	234.5	=	469
240-249	l l l l	x	244.5	=	978
250-259	l l l	x	254.5	=	763.5
260-269	l	x	265.5	=	265.5
	30				**6456**

$$\text{Weighted Average} = \frac{\text{Sum of the Products}}{\text{Total Frequency}}$$

$$= \frac{6456}{30}$$

$$= \boxed{215}$$

Now calculate the arithmetic mean without grouping data:

$$\text{Arithmetic Mean} = \frac{\text{Sum of All Measurements}}{\text{Number of Measurements Used}}$$

$$= \frac{6455}{30}$$

$$= \boxed{215}$$

The averages are the same. Remember, weighted averages using groupings are not normally done unless there is a much larger set of data to analyze.

The example given in this section illustrates the use of a weighted average calculation when data has been grouped. There is a more general equation for weighted average:

$$\bar{x} = \frac{w_1 x_1 + w_2 x_2 + ... + w_n x_n}{w_1 + w_2 + ... + w_n}$$

where:

$\bar{x}$ = arithmetic mean

w = the weight assigned to each number

x = each data point or number

This equation is used less frequently in water and wastewater calculations than the grouping equation shown in the examples.

---

* Often tic marks are used to mark off frequency as each data point is registered in the appropriate grouping.

PRACTICE PROBLEMS 6.3: Weighted Averages

❏ In problems 1 and 2, calculate the weighted average using 10-pt groups. How does this arithmetic mean compare with the one calculated without grouping?

1.

170	126	182	146	168	145	115	108
115	147	141	159	174	151	136	110
122	164	136	129	192	137	144	121
105	137	107	118	181	123	153	143
118	153	124	120	164	117	140	181

Weighted Average _____

Regular Arithmetic Mean _____
    (No grouping)

2.

240	241	210	247	211	219
207	217	212	264	235	227
219	222	231	258	228	249
236	245	217	219	248	266
250	261	225	208	233	240

Weighted Average _____

Regular Arithmetic Mean _____
    (No grouping)

7 *Ratios and Proportions*

Complete and score the following skills test. Each section should be scored separately in the box provided to the right. For Sections 7.1 and 7.2, a score of 8 or above indicates that you are sufficiently strong in that concept. A score of 7 or below indicates a review of that section is advisable. For Section 7.3, a score of 4 or above indicates that you are sufficiently strong in that concept. A score of 3 or below indicates a review of that section is advisable.

7.1 Determining Proportions

Number Correct

❑ Indicate whether the following ratios are proportionate. Write a "yes" or "no" in the answer blank. If yes, indicate the cross product.

1. 2:3 and 18:24

———

———

2. 1:3 and 9:27

———

———

3. 3:5 and 21:30

———

———

4. 45:72 and 5:9

———

———

5. 1:15 and 16:240

———

———

6. 28:220 and 84:640

———

———

7. 5:6 and 26:30

———

———

8. 27:30 and 18:20

———

———

9. 22:231 and 10:105

———

———

10. 6:7 and 114:126

———

———

7.2 Solving a Proportion

❑ Solve for the unknown value in each of the following proportions.

1. $1:5 :: x:110$

_____

2. $x:6 = 8:24$

_____

3. $\dfrac{3}{x} = \dfrac{12}{20}$

_____

4. $4/9 = 160/x$

_____

5. $4:850 = x:5100$

_____

6. $2/17 = x/306$

_____

7. $\dfrac{125}{25} = \dfrac{x}{2}$

_____

8. $\dfrac{28}{x} = \dfrac{525}{1125}$

_____

9. $21:7 :: x:4$

_____

10. $\dfrac{2}{x} = \dfrac{30}{105}$

_____

7.3 Setting Up A Proportion

❑ Set up and solve the proportion problems given below.

1. Four bags of chemical cost $130. At the same unit price, how much would 11 bags of chemical cost?

ANS_____

2. If 3 gallons of paint cover 950 sq ft, how many gallons of paint will be required to paint 2400 sq ft?

ANS_____

3. On the average, one barrel of chemical is used up in 17 days. At this rate, how many barrels will be required during a 90-day period?

ANS_____

4. One gallon is equivalent to 3.7853 liters. How many gallons are equivalent to 50 liters?

ANS_____

5. An average of 3 cubic feet of screenings are removed per million gallons of wastewater treated. At this rate, how many cubic feet of screenings would be expected to be removed from a flow of 4.6 MGD?

ANS_____

NOTES:

7.1 DETERMINING PROPORTIONS

SUMMARY

There are three methods to determine if two ratios are proportionate:

1. **Division** of each ratio.

$$\frac{1}{4} = \boxed{0.25} \qquad\qquad \frac{5}{20} = \boxed{0.25}$$

Since the answers are equal,
the <u>ratios are proportionate</u>:

2. **Means and extremes** of the ratios.

$$1:4 \ :: \ 5:20$$

$$\boxed{4 \times 5 = 20}$$

$$1 \times 20 = 20$$

Since the answers are equal the <u>ratios are proportionate</u>.

3. **Cross multiplication** of the ratios.

$$\frac{1}{4} = \frac{5}{20}$$

$$\boxed{(4)\,(5) = 20}$$

$$\boxed{(1)\,(20) = 20}$$

Since the answers are equal,
the <u>ratios are proportionate</u>:

WHAT ARE RATIOS AND PROPORTIONS?

A **ratio** is the established relationship between two numbers. For example, if an average of 2 cu ft of screenings are removed from each million gallons of wastewater treated, the ratio of screenings removed (cu ft) to treated wastewater (MG) is 2:1. As another example, the number of feet to each yard is a 3:1 ratio (3 feet to each 1 yard). Ratios are normally written using a colon (such as 2:1), or written as a fraction (such as 1/2).

A **proportion** exists when the value of one ratio is equal to the value of a second ratio. For example, the value of 1/2 is equal to the value of 4/8. Therefore, these two ratios are in proportion. Proportions may be written using colons or may be written using fractions. For example,

$$1:2 :: 4:8$$

is read as "one is to two as four is to 8." In other words, the value of the ratio 1:2 is the same as the value of 4:8. Written as fractions, the proportion is:

$$\frac{1}{2} = \frac{4}{8}$$

There are three methods that may be used to determine if two ratios are in proportion—division, means and extremes, and cross multiplication. Each of these methods is described in this section.

Example 1: (Determining Proportions)
❏ Use the division method to determine if the following two ratios are proportionate: 3/5 and 15/20.

First divide each ratio, then compare the answers:

$$\frac{3}{5} = \boxed{0.6} \qquad \frac{15}{20} = \boxed{0.75}$$

Since the answers are **not equal**, the ratios are **not in proportion**.

MEANS, EXTREMES AND CROSS MULTIPLICATION

If the proportion is written using colons, the product of the means will equal the product of the extremes. In other words, the inside terms multiplied together will equal the outside terms multiplied together:

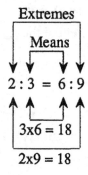

$$2 : 3 = 6 : 9$$

$$3 \times 6 = 18$$
$$2 \times 9 = 18$$

If the proportion is written using fractions, cross-multiplied terms will be equal:

$$\frac{2}{3} = \frac{6}{9} \qquad \begin{array}{l} 3 \times 6 = 18 \\ 2 \times 9 = 18 \end{array}$$

Example 2: (Determining Proportions)
❏ Use the means and extremes to determine if the following two ratios are proportionate: 24:4 and 18:3.

$$24 : 4 \;::\; 18 : 3$$

$$4 \times 18 = 72$$

$$24 \times 3 = 72$$

Since the product of the means is equal to the product of the extremes, the ratios are proportionate.

Example 3: (Determining Proportions)
❏ Use cross multiplication to determine if the ratios given below are proportionate: 4/5 and 72/95.

$$\frac{4}{5} = \frac{72}{95}$$

$$5 \times 72 = 360$$

$$4 \times 959 = 380$$

Since the products of the cross-multiplication are different, the ratios are not proportionate.

$$\frac{4}{5} \neq \frac{72}{95}$$

Example 4: (Determining Proportions)
❏ Use cross multiplication to determine if the following two ratios are proportionate: 12/9 and 132/99.

$$\frac{12}{9} = \frac{132}{99}$$

$$9 \times 132 = 1188$$

$$12 \times 99 = 1188$$

Since the cross multiplication products are equal, the two ratios are proportionate.

$$\frac{12}{9} = \frac{132}{99}$$

USING DIVISION

Because a proportion is **two equal ratios,** you can determine if two fractions are proportionate by simply dividing each fraction. If the answers are identical, the two fractions (or ratios) are proportionate. Example 1 illustrates this calculation.

USING MEANS AND EXTREMES

There are four terms to every proportion:

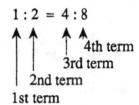

$$1 : 2 = 4 : 8$$

4th term
3rd term
2nd term
1st term

The outside terms (1st and 4th) are called the **extremes** of a proportion. The inside terms (2nd and 3rd) are called the **means** of a proportion.

For all proportions, the product of the means is equal to the product of the extremes. In other words, when the two inside terms are multiplied, the answer will be the same as when the two outside terms are multiplied.

If the products of the inside and outside terms are not equal, the two ratios are not proportionate. Example 2 illustrates a means and extremes calculation.

USING CROSS MULTIPLICATION

When two ratios are written as fractions, cross multiplication can be used to determine if the two ratios are proportionate. Examples 3 and 4 illustrate the use of cross multiplication.

PRACTICE PROBLEMS 7.1: DETERMINING PROPORTIONS

❑ Use the division method to determine if the following pairs of ratios are proportionate. If they are proportionate, give the quotient (or answer) of the division problem.

1. 5/6 and 70/84

———————

———————

3. 5/11 and 3900/8580

———————

———————

2. 7/4 and 574/320

———————

———————

❑ Use the means and extremes method to determine if the following pairs of ratios are proportionate. If they are proportionate, give the cross product.

4. 3:4 :: 576:770

———————

———————

6. 14:19 :: 42:57

———————

———————

5. 2:9 = 6:27

———————

———————

❑ Use the cross multiplication method to determine if the following pairs of ratios are proportionate. If they are proportionate, give the cross product.

7. $\dfrac{3}{8}$ and $\dfrac{87}{232}$

———————

———————

9. $\dfrac{4}{3}$ and $\dfrac{1760}{1350}$

———————

———————

8. $\dfrac{2}{3}$ and $\dfrac{234}{351}$

———————

———————

10. $\dfrac{6}{7}$ and $\dfrac{114}{173}$

———————

———————

7.2 SOLVING A PROPORTION PROBLEM

SUMMARY

To solve a proportion problem, use the same steps as solving for the unknown value:

1. All terms may only be moved **diagonally** from one side of the equation to the other.

2. To solve for x, first get x in the numerator.

3. Then get x by itself by moving all other terms away from x to the other side of the equation.

SOLVING A PROPORTION IS SOLVING FOR THE UNKNOWN VALUE

There are four terms in every proportion. In a proportion problem, three of the terms are known and one is unknown.

To solve for the unknown term, follow the steps solving for the unknown value:*

1. Get the x term in the numerator.

2. Get x by itself.

To accomplish these moves, remember that the **terms may only be moved diagonally** from one side of the equation to the other. For example, a term in the denominator may only be moved to the numerator of the other side of the equation.

WHEN TO USE CROSS MULTIPLICATION

When the x term is in the denominator, cross multiplication is often used as the first step in solving for the unknown value. Example 2 illustrates this type of problem.

Example 1: (Solving a Proportion Problem)
❑ Solve for x in the proportion problem given below.

$$\frac{26}{190} = \frac{x}{4750}$$

Since x is already in the numerator, to solve for x you merely need to get x by itself. Move the 4750 diagonally across to the other side of the equation, leaving x where it is:

$$\frac{(4750)\,(26)}{190} = x$$

$$\boxed{650} = x$$

Example 2: (Solving a Proportion Problem)
❑ Solve for the unknown value x in the problem given below.

$$\frac{3.2}{2} = \frac{6}{x}$$

First, cross multiply terms:

$$(3.2)\,(x) = (2)\,(6)$$

Now solve for the unknown value. Since x is already in the numerator, simply get x by itself:

$$x = \frac{(2)\,(6)}{3.2}$$

$$x = \boxed{3.75}$$

---

* Refer to Chapter 2, "Solving for the Unknown Value".

Example 3: (Solving a Proportion Problem)
❑ Given the proportion 5:9 :: x:72, solve for the unknown value.

First, rewrite the proportion in fraction form:

$$\frac{5}{9} = \frac{x}{72}$$

Then solve for the unknown value. Since x is already in the numerator, get x by itself by moving 72 diagonally to the other side of the equation:

$$\frac{(72)\,(5)}{9} = x$$

$$\boxed{40} = x$$

When the proportion is written horizontally, using colons, you may either rewrite the problem in fraction form and solve for the unknown value (as shown in Example 3) or multiply means and extremes first and then solve for the unknown value (as shown in Example 4).

Example 4: (Solving a Proportion Problem)
❑ Given the proportion 3:5 = 273:x, solve for x.

First multiply means and extremes of the proportion:

$$(5)\,(273) = (3)\,(x)$$

Now solve for the x term as usual:

$$\frac{(5)\,(273)}{3} = x$$

$$\boxed{455} = x$$

SHORTCUTS IN SOLVING PROPORTION PROBLEMS

Sometimes it is possible to solve proportion problems without ever lifting a pencil. The two shortcut methods most often used are:

- The **cross-multiplication** (or means and extremes) shortcut, and

- The **equivalent fractions** shortcut

When using the cross multiplication method, remember that **each cross multiplication is equal to the same number.** For example,

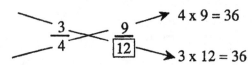

from cross multiplication we know that 2 x 15 will be the same answer as 3 x *x*. Therefore cross multiplication tells us that 3 times some number must equal 30. The unknown number must be 10 in this example since 3 x 10 = 30. To check:

$$\frac{2}{3} \times \frac{10}{15} \quad \begin{matrix} 3 \times 10 = 30 \\ 2 \times 15 = 30 \end{matrix}$$

For smaller fractions, this cross multiplication shortcut can be done mentally. Only larger fractions require this be done on paper. (See Example 5.)

You may use means and extremes values in the same way as cross multiplication. Example 6 illustrates this type of shortcut.

Example 5: (Solving a Proportion Problem)
❑ Use the cross multiplication method to determine the value of *x* in the following proportion:

$$\frac{3}{4} = \frac{9}{x}$$

The cross multiplication is equal to 36. Since 3 times *x* must equal 36, *x* must be equal to 12.

To check:

$$\frac{3}{4} \times \frac{9}{\boxed{12}} \quad \begin{matrix} 4 \times 9 = 36 \\ 3 \times 12 = 36 \end{matrix}$$

Example 6: (Solving a Proportion Problem)
❑ Use the means and extremes method to determine the value of *x* in the following proportion.

$$1{:}4 = x{:}20$$

The products of the means and extremes in this problem must both equal 20; therefore, *x* must be equal to 5:

$$1 : 4 = \boxed{5} : 20$$

$$4 \times 5 = 20$$
$$1 \times 20 = 20$$

Example 7: (Solving a Proportion Problem)
❏ Use the equivalent fractions method to determine the value of x in the following proportion:

$$\frac{7}{4} = \frac{28}{x}$$

To get from 7 to 28, you must multiply by 4. Therefore multiply both the numerator and denominator by 4:

$$\boxed{\frac{7}{4}} \begin{array}{l} \times 4 \\ \times 4 \end{array} = \boxed{\frac{28}{16}}$$

The value of x must therefore be 16.

Example 8: (Solving a Proportion Problem)
❏ Use the equivalent fractions method to determine the value of x in the following proportion:

$$\frac{63}{x} = \frac{9}{5}$$

To get from 63 to 9, you must divide by 7. Therefore divide both the numerator and denominator by 7:

$$\boxed{\frac{63}{x}} \begin{array}{l} \div 7 \\ \div 7 \end{array} = \boxed{\frac{9}{5}}$$

What number divided by 7 equals 5? The answer is 35. x must therefore be equal to 35.

A proportion is defined as two equal ratios. However, a proportion could also be thought of as two equivalent fractions.* In other words, the numerator and denominator of the first fraction can be multiplied or divided by some number in order to obtain the second fraction. Using the proportion 2:3 and 10:15, notice that multiplying the numerator and denominator of the first fraction by 5, results in the second fraction:

$$\boxed{\frac{2}{3}} \begin{array}{l} \times 5 \\ \times 5 \end{array} = \boxed{\frac{10}{15}}$$

Let's examine another proportion:

$$\frac{7}{9} = \frac{x}{81}$$

Using the concept of equivalent fractions, we will need to **multiply or divide the numerator and denominator of the first fraction by the same number** to obtain the numerator and denominator of the second fraction. In the example above, first examine the denominators. To go from 9 to 81, you will need to multiply by 9. Therefore multiply both the numerator and denominator by 9 to determine the value of x:

$$\boxed{\frac{7}{9}} \begin{array}{l} \times 9 \\ \times 9 \end{array} = \boxed{\frac{63}{81}}$$

The value of x is found to be 63. Examples 7 and 8 further illustrate the equivalent fraction shortcut.

---

* Refer to Chapter 3, Fractions, for a review of equivalent fractions.

PRACTICE PROBLEMS 7.2: Solving a Proportion Problem

❑ Solve for x in each problem given below. Use shortcut methods whenever possible.

1. $2{:}3 = 6{:}x$

$x = $ _____

6. $15{:}\ 3 :: x{:}4$

$x = $ _____

2. $25{:}x :: 10{:}2$

$x = $ _____

7. $x{:}\ 30 = 8{:}12$

$x = $ _____

3. $\dfrac{9}{3} = \dfrac{x}{8}$

$x = $ _____

8. $\dfrac{3}{8} = \dfrac{21}{x}$

$x = $ _____

4. $\dfrac{x}{27} = \dfrac{3}{9}$

$x = $ _____

9. $\dfrac{4}{x} = \dfrac{196}{1225}$

$x = $ _____

5. $1{:}\ 144 :: x{:}1296$

$x = $ _____

10. $\dfrac{x}{8} = \dfrac{49}{56}$

$x = $ _____

7.3 SETTING UP A PROPORTION PROBLEM

SUMMARY

Most water and wastewater problems involving proportions are **direct proportions**; that is, as one unit increases, the other increases as well. Occasionally (such as for some pump characteristics proportions), you may encounter a problem that involves an **inverse proportion** (as one unit increases, the other decreases). The set up and solution of direct and inverse proportions are as follows:

To set up and solve direct proportions:

1. Write the two fractions, being careful that the **location of the units is the same for each fraction.**

2. Fill in the given values for both fractions (x will be one of the values for the fraction on the right side).

3. Solve for the unknown value.

To set up and solve indirect proportions:

1. Group like units.

2. Place the smaller numbers in the numerators and the larger numbers in the denominators.

3. Solve for the unknown value.

As shown in the diagram to the right, in a **direct proportion**, the two units of the problem change in the **same direction**. For example, if gallons increase, square feet increase or, if gallons decrease, square feet decrease.

In an **inverse proportion**, the two units of the problem change in **opposite directions**. This means as one unit increases, the other decreases, and vice versa.

DIRECT PROPORTIONS

Most proportion problems in water and wastewater math are <u>direct proportions</u> problems. For example, conversion calculations such as converting cubic feet to gallons may be done using proportions; and various chemical calculations may be made using proportions. The first setp in setting up direct proportion problems is simply writing the two fractrions. For example, if the two types of units given in the problem are gallons and square feet, the units of the proportion would be written as:

$$\frac{gal}{sq\ ft} = \frac{gal}{sq\ ft}$$

The second step in setting up a direct proportion is simply filling in the given information. The fraction to the left will always have both numbers filled in and the fraction to the right will always contain the *x* term. For example:

$$\frac{1\ gal}{300\ sq\ ft} = \frac{x\ gal}{300\ sq\ ft}$$

Once the proportion has been set up, then solve for *x*.* Examples 1 and 2 illustrate this calculation.

DIRECT PROPORTIONS
CHANGE IN THE SAME DIRECTION

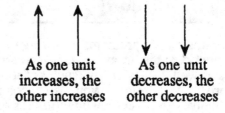

| As one unit increases, the other increases | As one unit decreases, the other decreases |

INVERSE PROPORTIONS
CHANGE IN OPPOSITE DIRECTIONS

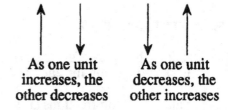

| As one unit increases, the other decreases | As one unit decreases, the other increases |

Example 1:   (Setting up a Proportion Problem)
❏ 0.5 lbs of chlorine are dissolved in 45 gallons of water. To maintain the same concentration, how many pounds of chlorine would have to be dissolved in 100 gallons of water?

First, write the two fractions:

$$\frac{lbs}{gallons} = \frac{lbs}{gallons}$$

Next, fill in the given information in the fractions:

$$\frac{0.5\ lbs}{45\ gal} = \frac{x\ lbs}{100\ gal}$$

Now solve for *x* as described in Section 7.2:

$$(45)\,(x) = (0.5)\,(100)$$

$$x = \frac{(0.5)\,(100)}{45}$$

$$= \boxed{1.1\ lbs}$$

---

* Refer to Chapter 2 for a review of solving for the unknown value.

Example 2: (Setting up a Proportion Problem)
❏ In the seeding of a new digester, for every 0.05 lbs of volatile solids entering the digester, 1 lb of volatile solids should be in the digester (in the seed sludge). If 150 lbs of volatile solids enter the digester daily, how many pounds of volatile solids should be under digestion?

Although both units in this problem are pounds, they describe two different types of pounds. First group like units:

$$\frac{\text{lbs VS entering}}{\text{lbs VS under digestion}} = \frac{\text{lbs VS entering}}{\text{lbs VS under digestion}}$$

Now fill in the numbers. The original ratio given is 0.05 lbs VS entering per 1 lb VS in the digester. Because more volatile solids are entering, more lbs of volatile solids will be needed in the digester:

$$\frac{0.05\ \text{lbs VS entering}}{1\ \text{lbs VS under digestion}} = \frac{150\ \text{lbs VS entering}}{x\ \text{lbs VS under digestion}}$$

Now solve for x:

$$(0.05)(x) = (150)(1)$$
$$x = \frac{(150)(1)}{0.05}$$
$$x = \boxed{3000\ \text{lbs VS under digestion}}$$

Example 3: (Setting up a Proportion Problem)
❏ It takes 3 men 60 hours to complete a job. At the same rate, how many hours would it take 5 men to complete the job?

This is an inverse proportion. As the number of men increases, the number of hours required to complete the job decreases. Therefore, the unknown hours should be less than 60 hours, and x is placed in the numerator:

$$\frac{3\ \text{men}}{5\ \text{men}} = \frac{x\ \text{hours}}{60\ \text{hours}}$$

Now solve for x:

$$\frac{(3)(60)}{5} = x$$
$$\boxed{36\ \text{hours}} = x$$

INVERSE PROPORTIONS

There are three basic steps in setting up and solving inverse proportions:

1. Group like units.

2. Place the smaller numbers in the numerators and the larger numbers in the denominators.

3. Solve for the unknown value.

The first step in setting up inverse proportion problems is **grouping like terms** into two fractions. For example, if the two types of units given in the problem are gallons and square feet, the two numbers representing gallons are placed in one fraction and the two numbers representing square feet are placed in the other fraction.

The second step in setting up proportions involves **placing the numbers in the proper position** in the fractions. The important point here is that the smaller numbers in each fraction be placed in the **same position**—either both in the numerator or both in the denominator; and similarly, that the larger numbers be placed in the same position in each fraction. For consistency, it is suggested that you always place the smaller numbers in the numerators of each fraction and the larger numbers in the denominators. Generally you will know by the context of the problem and experience whether the x will be larger or smaller than the other number with the same units.

In the third step, you will solve for x just as for direct proportions. Example 3 illustrates this type of calculation.

PRACTICE PROBLEMS 7.3: Setting Up A Proportion

❑ Set up and solve the following proportion problems.

1. One gallon is equivalent to 3.7853 liters. How many gallons are equivalent to 75 liters?

   ANS_____

2. On the average one bag of chemical is used up in 3.5 days. At this rate, how many bags of chemical will be required during a 120-day period?

   ANS_____

3. Suppose you wish to maintain a weir overflow rate of 12,000 gpd/ft. (This is 12,000 gpd flow for each 1 ft of weir length.) If the weir length is 180 ft, what gpd flow will result in the desired weir overflow rate?

   ANS_____

4. A total of 5.4 lbs of hypochlorite are dissolved in 80 gallons of water. For a solution with the same concentration, how many lbs of hypochlorite must be dissolved in 30 gallons of water?

   ANS_____

5. A treatment pond is designed for a population loading of 300 persons per acre of pond. If the population to be served is 1240 people, how many acres of treatment pond will be required?

   ANS_____

8 *Conversions*

SKILLS CHECK

Complete and score the following skills test. Each section should be scored separately in the box provided to the right. A score of 4 or above indicates you are sufficiently strong in that concept. A score of 3 or below indicates a review of that section is advisable.

8.2 Cubic Feet to Pounds Conversions*

Number
Correct

❑ Complete the conversion calculations indicated below. (The conversion equations include 1 cu ft = 7.48 gal and 1 gal = 8.34 lbs.)

1. 48,000 cu ft = _____ gal

4. 120 cu ft = _____ lbs

2. 310,000 lbs = _____ cu ft

5. 81,400 lbs = _____ gal

3. 186,000 gal = _____ lbs

8.3 Flow Conversions

Number
Correct

❑ Complete the following conversion calculations. (Use the conversion equations given in Section 8.2 above.)

1. 2.5 cfs = _____ gpm

4. 1500 gpm = _____ cfs

2. 2460 gpm = _____ cfm

5. 3.52 MGD = _____ cfs

3. 2,195,000 gpd = _____ gpm

* The skills check sections have been numbered to correspond with the sections in this chapter. Section 8.1 is an introduction to the "box method" of conversions.

8.4 Linear Measurement Conversions

❏ Complete the following linear measurement conversions. (The conversion equations include 1 ft = 12 in., 1 yd = 3 ft, and 1 mi = 5,280 ft.)

1.   28 in. = _____ ft

4.   1800 ft = _____ mi

2.   4.7 mi = _____ ft

5.   260 ft = _____ yds

3.   3.7 ft = _____ in.

8.5 Area Measurement Conversions

❏ Complete the conversions indicated below. (Use conversion equations adapted from those shown in Section 8.4. In addition, 1 ac = 43,560 sq ft.)

1.   1640 sq ft = _____ sq yds

4.   135,000 sq ft = _____ ac

2.   4.8 acres = _____ sq ft

5.   0.46 ac = _____ sq ft

3.   2100 sq in = _____ sq ft

8.6 Volume Measurement Conversions

Number Correct

❑ Complete the following volume measurement conversions. (Conversion equations include 1 cu ft = 7.48 gal, 1 ac-ft = 43,560 cu ft.)

1. 35,300 cu ft = _____gal

4. 1.75 ac-ft = _____cu ft

2. 42 cu yds = _____cu ft

5. 420,000 cu ft = _____ac-ft

3. 128,355 cu in = _____cu ft

8.7 mg/*L* and Percent Conversions

Number Correct

❑ Complete the following conversions.

1. 320 mg/*L* = _____%

4. 210 mg/*L*= _____%

2. 1.2% = _____mg/*L*

5. 0.26% = _____mg/*L*

3. 0.01% = _____mg/*L*

8.8 mg/*L* and gpg Conversions

❑ Complete the conversions below. (Use the conversion equation 1 gpg = 17.1 mg/*L*.)

1.   12 gpg = _____mg/*L*

4.   110 mg/*L* = _____gpg

2.   9.1 gpg = _____mg/*L*

5.   14.2 gpg = _____mg/*L*

3.   195 mg/*L* = _____gpg

8.9 Metric System Conversions (Metric to Metric)

❑ Complete the metric system conversions given below.

1.   145 mg = _____g

4.   520 ml = _____*L*

2.   2.5 *L* = _____ml

5.   12 g = _____mg

3.   200 mm = _____m

8.10 Metric/English Conversions

❏ Complete the metric/English conversions shown below using the conversion factors listed below.

1. 5,480 lbs = _____kg

4. 6.26 MGD = _____m^3/d

2. 50 gal = _____ *L*

5. 6.8 m = _____ft

3. 1.7 m^3/s = _____cfs

1 MGD	=	3785 m^3/d
1 cfs	=	0.0283 m^3/s
1 gal	=	3.785 *L*
1 lb	=	0.4536 kg
1 ft	=	0.3048 m

NOTES:

8.1 THE "BOX METHOD" OF CONVERSIONS

SUMMARY

There are three basic considerations when using the box method of conversions:

1. When moving from a smaller box to a larger box, multiplication is indicated.

2. When moving from a larger box to a smaller box, division is indicated.

3. To use the "box method" successfully, be sure the conversion equation is in the form of the example shown below:

$$1 \text{ yd } = 3 \text{ ft}$$

↑ ↑

Always have a 1 on the left side of the equation.   *Always have a number **greater than 1** on the right side of the equation.*

If it is not in this format, recalculate the equation so that it is in the proper form.

Many times, when making conversions from one unit to another, people become confused about whether to multiply or divide by the conversion factor. The **box method of conversions** was developed as an aid in making that decision. Although there are many ways of approaching conversions, this method has been singled out because it seems to be useful to a wider audience and results in fewer errors. You may wish to use some other method of conversion calculation.

To use the box method of conversions, first set up the boxes. The conversion equation should have the number one on the left side of the equation and a number greater than one on the right side of the equation. Therefore, the box on the left will always be smaller and the box on the right will always be larger.

Multiplication: Because multiplication* is associated with **increasing** a number, we will use multiplication when moving from a smaller box to a larger box.

Division: Because division* is associated with **decreasing** a number, we will use division when moving from a larger box to a smaller box.

HOW TO SET UP THE BOXES

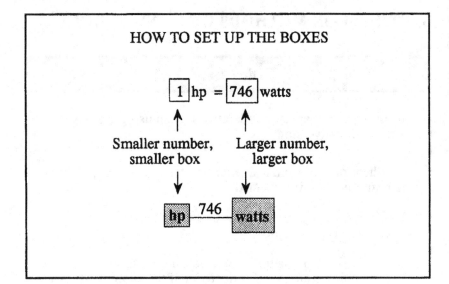

MULTIPLICATION AND DIVISION

Moving from the smaller box to the larger box, multiplication is indicated:

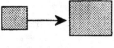

Multiply

Moving from the larger box to the smaller box, division is indicated:

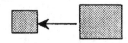

Divide

---

* Using numbers greater than 1.

Example 1: (Box Method of Conversions)
❑ 12 kw is equivalent to how many horsepower?
(1 kw = 1.341 hp)

First, draw the box diagram:

$$\boxed{\text{kw}} \xrightarrow{1.341} \boxed{\text{hp}}$$

Converting from kilowatts to horsepower, we are moving from a smaller box to a larger box; therefore multiplication is indicated:

$$(12 \text{ kw}) (1.341 \text{ hp/kw}) = \boxed{16 \text{ hp}}$$

Example 3: (Box Method of Conversions)
❑ Express 5.1 feet in terms of meters. (1 ft = 0.3048 m).

The conversion equation is not in the desired form. The desired form can be obtained by dividing both sides of the equation by the decimal fraction:

$$\frac{1 \text{ ft}}{0.3048} = \frac{0.3048 \text{ m}}{0.3048}$$

$$3.28 \text{ ft} = 1 \text{ m}$$

The "1" is usually written on the left side of the equation:

$$1 \text{ m} = 3.28 \text{ ft}$$

$$\boxed{\text{m}} \xleftarrow{3.28} \boxed{\text{ft}}$$

Now complete the conversion using the box method. From feet to meters, division is indicated:

$$\frac{5.1 \text{ ft}}{3.28 \frac{\text{ft}}{\text{m}}} = \boxed{1.55 \text{ m}}$$

WHEN THE CONVERSION EQUATION INCLUDES A DECIMAL FRACTION

The box method cannot be used when the conversion equation includes a decimal fraction (a decimal number less than one).*

However, if you wish to use the box method, simply convert the equation to the proper form by **dividing both sides of the equation by the decimal fraction**. Example 2 illustrates such a calculation.

USING DIMENSIONAL ANALYSIS

Another method commonly used to determine whether multiplication or division is to be used is dimensional analysis. Refer to Chapter 15 for a discussion of dimensional analysis.

---

* The box method is based on the concept that when a number is multiplied by *a number greater than one*, the answer is larger than the original number, and when divided the answer is smaller. Neither of these conditions is true when multiplying or dividing by a decimal fraction.

PRACTICE PROBLEMS 8.1: The "Box Method" of Conversions

❑ Complete the following problems using the box method of conversion. Use the conversion equations given below.

1. 32 in. =_____cm

6. 105 yds =_____m

2. 120 cu ft =_____m^3

7. 18 kw =_____hp

3. 37 m =_____yds

8. 515 lbs =_____kg

4. 75 kg =_____lbs

9. 100 cm =_____in.

5. 250 ml =_____oz (fl.)

10. 30 hp =_____kW

1 cm = 2.54 in.
1 hp = 0.746 kW
1 m^3 = 35.31 cu ft
1 lb = 0.4536 kg
1 yd = 0.9144 m
1 oz (fluid) = 29.57 ml

8.2 CUBIC FEET TO POUNDS CONVERSIONS

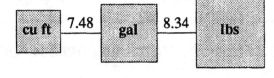
---

* For a discussion of cubic inches, cubic feet, and cubic yards conversions, refer to Section 8.6.
** Refer to Section 8.1 for a discussion of the "box method" of conversions.

The box method may be used when converting from cubic feet to gallons and from gallons to pounds. The diagram to be used in making these conversions is given at the bottom of the page.

Remember, when moving **from a smaller box to a larger box, use multiplication:**

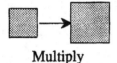

Multiply

And when moving from a **larger box to a smaller box, use division:**

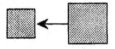

Divide

Examples 1 and 2 illustrate cu ft/gal/lbs conversion problems.

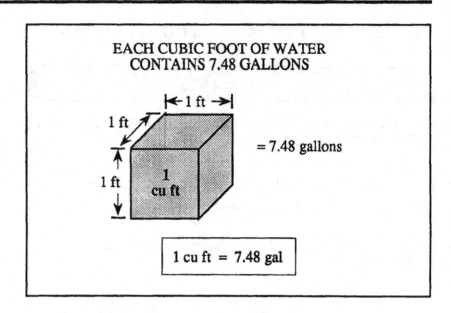

EACH CUBIC FOOT OF WATER CONTAINS 7.48 GALLONS

= 7.48 gallons

1 cu ft = 7.48 gal

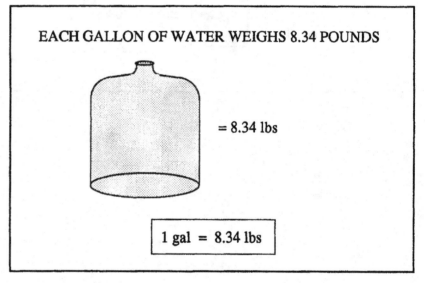

EACH GALLON OF WATER WEIGHS 8.34 POUNDS

= 8.34 lbs

1 gal = 8.34 lbs

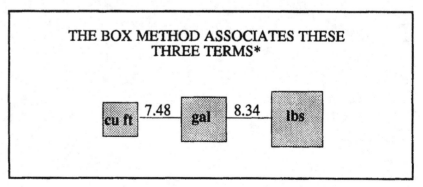

THE BOX METHOD ASSOCIATES THESE THREE TERMS*

cu ft —7.48— gal —8.34— lbs

---

* To convert directly from cubic feet to pounds, use 62.4 lbs/cu ft. When the factors 7.48 and 8.34 (equal to 62.38) are used for this conversion, the resulting answer will be slightly different.

Example 1: (Cu ft to lbs Conversions)

❑ A tank has a capacity of 60,000 cu ft. What is the gallon capacity of the tank?

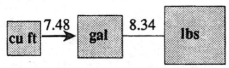

We are beginning with cubic feet (60,000 cu ft) and wish to convert to gallons. Therefore we are moving from a smaller box to a larger box and multiplication is indicated:

$$(60,000 \text{ cu ft}) \left(7.48 \frac{\text{gal}}{\text{cu ft}}\right) = \boxed{448,800 \text{ gal}}$$

Example 2: (Cu ft to lbs Conversions)

❑ 550,000 lbs of digested sludge are to be sent to the drying beds. Assuming each gallon of digested sludge weighs 8.34 lbs*, how many cu ft of sludge will be sent to the drying beds?

Converting from lbs to cu ft, we are moving from a larger box to smaller boxes; therefore, division is indicated:

$$\frac{550,000 \text{ lbs}}{\left(7.48 \frac{\text{gal}}{\text{cu ft}}\right)\left(8.34 \frac{\text{lbs}}{\text{gal}}\right)} = \boxed{\begin{array}{l}8816 \text{ cu ft sludge} \\ \text{to drying beds}\end{array}}$$

PLACING THE CONVERSION FACTORS

When division is indicated, always place that factor in the denominator. And when multiplication is indicated, place that factor in the numerator. In Example 2, the conversion from lbs to gal indicated division, as did the conversion from gal to cu ft. Both factors were therefore placed in the denominator of the problem.

It is best to calculate the problem all in one step rather than making several separate calculations. This reduces the chance of error. Using most simple calculators, Example 2 would be calculated as follows:

• Enter 550,000,

• Press the ÷ key and enter the first divisor (7.48),

• Then press the ÷ key again and enter the second divisor (8.34),

• Press the equal sign (=).

---

* Many times sludge weighs more than 8.34 lbs/gal. This can be verified by weighing a gallon of "typical" sludge. Refer to Chapter 7, Pumping Problems, in the applied math texts.

PRACTICE PROBLEMS 8.2: Cubic Feet to Pounds Conversions

❑ Complete the conversion indicated below.

1. 4 cu ft =_____gal

2. 1000 gal =_____cu ft

3. 35 cu ft =_____lbs

4. 60,000 gal =_____lbs

5. 1500 lbs =_____gal

❑ Solve the following conversion problems.

6. A tank contains 188,000 gallons. How many cu ft is this?

ANS_____

7. Convert 60,000 lbs of sludge to cu ft of sludge. (Assume the sludge weighs 8.34 lbs/gal)

ANS_____

8. If a tank contains 107,000 lbs of water, how many gallons of water does it contain?

ANS_____

9. How much does 20 cu ft of water weigh (in lbs)?

ANS_____

10. A tank contains 500 gallons of water. How many lbs of water does it contain?

ANS_____

8.3 FLOW CONVERSIONS

SUMMARY

To convert from one flow unit to another, several conversion equations would be required. The box method of flow conversions incorporates 12 conversion equations and many more conversion possibilities into one diagram:

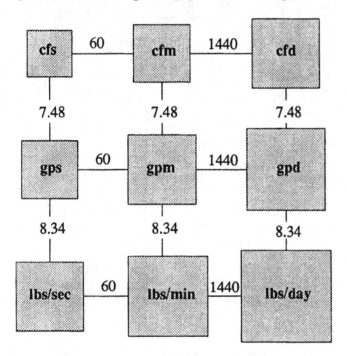

Two conversion equations which are commonly used but not shown directly on this diagram* are:

$$1 \text{ MGD} = 1.55 \text{ cfs}$$

$$1 \text{ MGD} = 694 \text{ gpm}$$

---

* Beginning with 1,000,000 gpd (1 MGD), each of these equations can be calculated using the box diagram.

The box method of conversions is especially useful in making flow conversions. Many of the possible conversions are visible in one graphic, as shown to the right.

When you become familiar with how this graphic is constructed, you will be able to construct it (or any part of it) any time the need arises.

First, look at the top row (cfs, cfm, cfd). Notice that **the only difference in these units is the time element**. The smallest time element (and smallest box) is to the left. And largest time element (and the largest box) is to the right. The number that connects seconds and minutes is 60; the number that connects minutes and day is 1440. Rows two and three are the same—only the time element changes.

Now let's look at the columns. The column to the far left begins with cfs (smallest box), then goes to gps (larger box) and ends with lbs/sec (even larger box). Notice that the time element (seconds) remains constant within the column, and the only change is cu ft to gal to lbs. The number that connects cu ft and gal is 7.48, and the number that connects gal and lbs is 8.34. The structure of the next two columns is the same.

As a general observation, notice that **the boxes get larger to the right and larger toward the bottom**. The smallest box is in the top left corner while the largest box is in the bottom right.

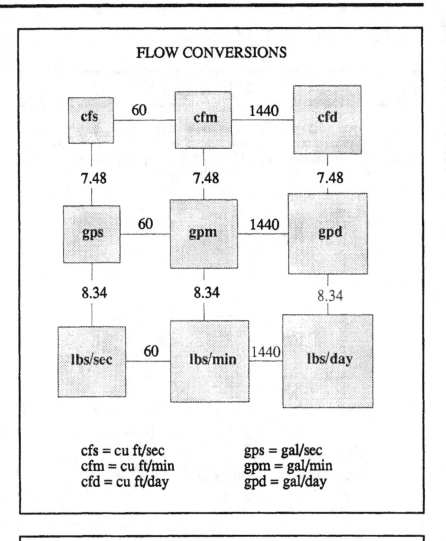

FLOW CONVERSIONS

cfs = cu ft/sec   gps = gal/sec
cfm = cu ft/min   gpm = gal/min
cfd = cu ft/day   gpd = gal/day

Example 1:   (Flow Conversions)
❏ If the flow in a pipeline is 2.3 cfs, what is this flow in gpm?

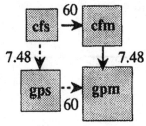

There are two possible paths from cfs to gpm. Each path has factors of 60 and 7.48, with only a difference in order. In each case the movement is from a smaller box to a larger box; therefore, multiplication is indicated:

$$(2.3 \; \underline{\text{cu ft}}) \; (60 \; \underline{\text{sec}}) \; (7.48 \; \underline{\text{gal}}) = \boxed{1032 \; \underline{\text{gal}}}$$
$$ \text{sec} \text{min} \text{cu ft} \text{min}$$

Example 2: (Flow Conversions)
❏ The flow to a treatment plant is 2,450,000 gpd. At this rate, what is the average cfs flow?

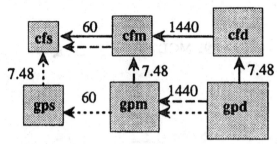

There are **three possible paths** from gpd to cfs. In each case, we are moving from a larger box to a smaller box, thus indicating division by 7.48, 1440, and 60 (in any order):

$$\frac{2,450,000 \ \frac{gal}{day}}{(7.48 \ \frac{gal}{cu \ ft})(1440 \ \frac{min}{day})(60 \ \frac{min}{sec})} = \boxed{3.8 \ \frac{cu \ ft}{sec}}$$

Example 3: (Flow Conversions)
❏ A flow of 4.61 MGD is equivalent to a flow of how many cfs?

We can use the conversion equation 1 MGD = 1.55 cfs. If desired, you may put this information in a box diagram:*

$$\boxed{1} \ MGD \ = \ \boxed{1.55} \ cfs$$

smaller number,   larger number,
smaller box   larger box

MGD $\xrightarrow{1.55}$ cfs

$$(4.61 \ MGD) (1.55 \ \frac{cfs}{MGD}) = \boxed{7.15 \ cfs}$$

*The set-up for box diagrams is described in Section 8.1.

It is always prudent to check your calculations. When using dimensional analysis as a check of the calculation set-up, you will need to write the units in expanded form. For example, cfs must be written as cu ft/sec and gpd must be written as gal/day. Refer to Chapter 15 for a discussion of dimensional analysis and its use in checking calculations.

CONVERSIONS THAT INCLUDE MGD

The box diagram does not include million gallons per day (MGD) since MGD can be so quickly converted to gpd (which *is* on the box diagram).

To convert from MGD to gpd, make the decimal point the millions comma:

2.87 MGD
↕
2,870,000 gpd

To convert from gpd to MGD, use the procedure in reverse—make the millions comma a decimal point:

1,710,000 gpd
↕
1.71 MGD

There are two conversion equations used quite frequently in water and wastewater treatment. You would be well advised to memorize these equations for use in quick conversions:

> 1 MGD = 1.55 cfs
> 1 MGD = 694 gpm

Should you forget these numbers, you can always derive them yourself using the box method of conversions:

$$\frac{1,000,000 \ \frac{gal}{day}}{(7.48 \ \frac{gal}{cu \ ft})(1440 \ \frac{min}{day})(60 \ \frac{min}{sec})} = \boxed{1.55 \ cfs}$$

$$\frac{1,000,000 \ \frac{gal}{day}}{(1440 \ \frac{min}{day})} = \boxed{694 \ gpm}$$

PRACTICE PROBLEMS 8.3: Flow Conversions

❑ Complete the following conversion calculations:

1. 3.6 cfs =_____gpm

5. 2.92 MGD =_____gpm

2. 1820 gpm =_____gpd

6. 385 cfm =_____gpd

3. 45 gps =_____cfs

7. 1,662,000 gpd =_____gpm

4. 8.6 MGD =_____gpm

8. 3.77 cfs =_____MGD

9. The flow through a pipeline is 2.8 cfs. What is this flow in gpd?

ANS_____

10. A treatment plant receives a flow of 3.61 MGD. What is this flow in gpm?

ANS_____

8.4 LINEAR MEASUREMENT CONVERSIONS

SUMMARY

To convert linear area measurements, the following conversion equations are most often used:

> 1 yard = 3 feet
> 1 foot = 12 inches

The "box method"* diagram that is associated with these measurements is:

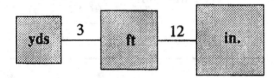

In certain aspects of water and wastewater treatment, another conversion equation is required:

> 1 mile = 5,280 ft

The corresponding box diagram is:

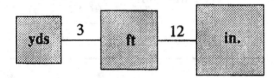

---

* Refer to Section 8.1 for a discussion of the "box method" of conversions.

YARDS, FEET, AND INCHES CONVERSIONS

Linear measurement, as the name suggests, is measurement along a line. It is often expressed in terms of yards, feet, or inches but can also be expressed as miles.

Many people find linear measurement conversions one of the easiest type of conversion calculations. Perhaps this is because these conversions are required so frequently in day-to-day circumstances. However, in a "crunch" situation such as a certification exam, it is a common experience for people to become confused when making these calculations.

Always remember that the simplest of equations can be diagramed using the "box method" of conversions, helping you decide whether to multiply or divide. The following two equations are used most often in linear measurement conversions:

$$1 \text{ yd} = 3 \text{ ft}$$
$$1 \text{ ft} = 12 \text{ in.}$$

These equations can be combined into one box diagram:

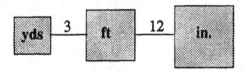

Examples 1 and 2 illustrate these conversion calculations.

Example 1: (Linear Measurement Conversions)
❏ The length of a fence is 162 yards. How many feet is this?

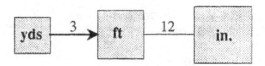

From yards to feet, we are moving from a smaller box to a larger box. Therefore multiplication is indicated:

$$(162 \text{ yds}) \frac{(3 \text{ ft})}{\text{yds}} = \boxed{486 \text{ ft}}$$

Example 2: (Linear Measurement Conversions)
❏ The total weir length of a sedimentation tank is 195 ft 10 inches. Express the length in ft only.

To express the length as feet only, the 10 inches must first be converted to feet:

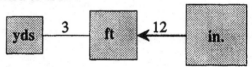

Moving from a larger box to a smaller box indicates division:

$$\frac{10 \text{ in.}}{\frac{12 \text{ in.}}{\text{ft}}} = 0.83 \text{ ft}$$

The 195 ft and 0.83 ft must now be combined:

$$\begin{array}{r} 195.0 \text{ ft} \\ + \underline{0.83 \text{ ft}} \\ 195.83 \text{ ft} \end{array}$$

Example 3: (Linear Measurement Conversions)
❑ A quarter-mile section of pipeline is to be isolated. How many feet of pipeline will be isolated?

A quarter mile is 1/4 mile or 0.25 mile. To convert to feet, use the box diagram:

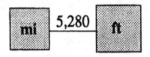

Moving from a smaller box to a larger box, multiplication is indicated:

$$(0.25 \text{ mi}) \left(5280 \frac{\text{ft}}{\text{mi}}\right) = \boxed{1320 \text{ ft}}$$

If you had used the fraction (1/4 mi), the calculation would still be 1/4 x 5280:*

$$\frac{(1 \text{ mile})}{4} \frac{(5280 \text{ ft})}{\text{mile}}$$

$$= \frac{5280}{4}$$

$$= \boxed{1320 \text{ ft}}$$

MILES AND FEET CONVERSIONS

In certain aspects of water and wastewater treatment, such as collection and distribution, conversions between feet and miles may be required. The conversion equation to be used is:

$$\boxed{1 \text{ mi } = 5,280 \text{ ft}}$$

And the corresponding box diagram is:

Examples 3 and 4 illustrate miles and feet conversion calculations.

Example 3: (Linear Measurement Conversions)
❑ A pipeline 4200 ft long is how many miles long? (Round to the nearest tenth mile.)**

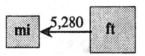

Division by 5280 is indicated:

$$\frac{4200 \text{ ft}}{5280 \frac{\text{ft}}{\text{mi}}} = \boxed{0.8 \text{ mi}}$$

---

* For a discussion of multiplication by fractions, refer to Chapter 3.
** For a review of rounding, refer to Chapter 14.

PRACTICE PROBLEMS 8.4: Linear Measurement Conversions

❏ Complete the following linear measurement conversion calculations:

1. 17 ft =_____yds

5. 492 in. =_____yds

2. 122 in. =_____ft

6. 0.6 ft =_____in.

3. 1.7 mi =_____ft

7. 70 yds =_____ft

4. 30 yds =_____in.

8. 9000 ft =_____mi

9. The total weir length for a sedimentation tank is 142 ft 7 in. Express this length in terms of feet only.

ANS_____

10. A one-eighth mile segment of pipeline is to be repaired. How many feet of pipeline is this?

ANS_____

8.5 AREA MEASUREMENT CONVERSIONS

SUMMARY

Area measurement conversion equations for square yards, square feet, and square inches are:

> 1 sq yd = 9 sq ft
>
> 1 sq ft = 144 sq in.

A single box diagram can be used for both of these conversion equations

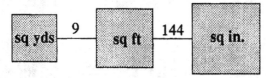

Another commonly used area measurement is acres. The conversion equation normally associated with acres is:

> 1 ac = 43,560 sq ft

And the box diagram is:

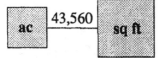

SQUARE TERMS CONVERSIONS

Square yards, square feet and square inches are all measurements of area. There are two ways to abbreviate these terms:

- sq yds, sq ft, sq in., and

- yd^2, ft^2, $in.^2$

The abbreviations using exponents* are useful because they indicate how many times the factors are to be multiplied—twice.

The factors used in making linear measurement conversions are shown in the box to the right. Notice that the **same factors** are used in area measurement conversions; however, they are **multiplied together twice** because the units have been squared.

Should you ever forget what numbers to use in sq yds, sq ft, and sq in. conversions, you can derive the equations for yourself. For example, using feet and inches conversions:

$$1 \text{ ft} = 12 \text{ in.}$$

Simply square both sides of the equation:

$$(1 \text{ ft})^2 = (12 \text{ in.})^2$$

$$(1 \text{ ft})(1 \text{ ft}) = (12 \text{ in.})(12 \text{ in.})$$

$$1 \text{ ft}^2 = \boxed{144 \text{ in.}^2}$$

Expressed as a box diagram,** this is:

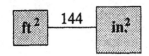

COMPARING LINEAR AND SQUARE TERMS

Linear Terms

yds —3— ft —12— in.

Square Terms

sq yds —(3) (3)— sq ft —(12) (12)— sq in.

sq yds —9— sq ft —144— sq in.

Example 1: (Area Measurement Conversions)
❑ The cross-sectional area of a pipe is 50.2 sq in.. How many sq ft is this?

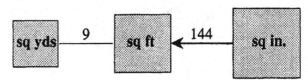

When converting from sq in. to sq ft, we are moving from a larger box to a smaller box; therefore, division is indicated:

$$\frac{50.2 \text{ sq in.}}{144 \text{ sq in./sq ft}} = \boxed{0.35 \text{ sq ft}}$$

---

* For a review of exponents, refer to Chapter 13, Powers, Roots, and Scientific Notation.
** Section 8.1 describes how to set up box diagrams.

Example 2: (Area Measurement Conversions)
❏ 200 sq yds is equivalent to how many sq ft?

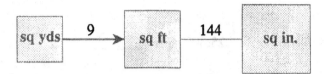

Converting from sq yds to sq ft, multiplication is indicated:

$$(200 \text{ sq yds}) \left(9 \frac{\text{sq ft}}{\text{sq yd}}\right) = \boxed{1800 \text{ sq ft}}$$

Example 3: (Area Measurement Conversions)
❏ A treatment plant requires an additional 0.75 acres for drying beds. How many sq ft are required?

From acres to sq ft, multiplication by 43,560 is indicated:

$$(0.75 \text{ ac}) \left(43,560 \frac{\text{sq ft}}{\text{ac}}\right) = \boxed{32,670 \text{ sq ft}}$$

ACRES AND FEET CONVERSIONS

Acres is also a measurement of area. Each acre is comprised of 43,560 sq ft. Thus, the conversion equation used in these calculations is:

$$\boxed{1 \text{ acre} = 43,560 \text{ sq ft}}$$

And the corresponding box diagram is:

Example 3 illustrates a calculation of this type.

---

* The set-up for box diagrams is described in Section 8.1.

PRACTICE PROBLEMS 8.5: Area Measurement Conversions

❏ Complete the square term conversion problems given below.

1. 1017 sq in. =_____sq ft

5. 25,000 sq ft =_____ac

2. 500 sq yds =_____sq ft

6. 1 sq yd =_____sq in.

3. 4 ac =_____sq ft

7. 9.5 sq ft =_____sq in.

4. 78.5 sq in. =_____sq ft

8. 0.9 ac =_____sq ft

9. For solids treatment, a total of 60,000 sq ft will be required. How many acres is this?

ANS_____

10. A pipe has a cross-sectional area of 452 sq in. How many sq ft is this?

ANS_____

8.6 VOLUME MEASUREMENT CONVERSIONS

SUMMARY

Volume measurement conversions of cubic yards, cubic feet and cubic inches involve the following two conversion equations:

> 1 cu yd = 27 sq ft
>
> 1 cu ft = 1728 cu in.

These equations may be expressed in a box diagram, as follows:

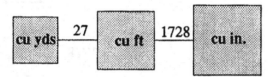

Another volume conversion used in water and wastewater calculations is:

> 1 ac-ft = 43,560 cu ft

This can be represented using the box method:

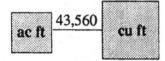

CUBIC TERMS CONVERSIONS

Volume measurements are generally expressed in cubic yards, cubic feet, and cubic inches. These terms may be abbreviated as:

- cu yds, cu ft, cu in., and

- yd^3, ft^3, $in.^3$

The factors used in making linear measurement conversions are shown in the diagram to the right—3 and 12. Note that the same factors (3 and 12) are used for cubic terms; however, **they are multiplied three times** because the units have been cubed.*

You may derive the cubic terms conversion equations by beginning with the linear terms equation, then cubing each side of the equation. For example, if you forgot the equation for cu ft and cu yds, you would first begin with the ft and yds equation:

$$1 \text{ yd} = 3 \text{ ft}$$

Then cube each side of the equation:

$$(1 \text{ yd})^3 = (3 \text{ ft})^3$$

$$(1 \text{ yd})(1 \text{ yd})(1 \text{ yd}) = (3 \text{ ft})(3 \text{ ft})(3 \text{ ft})$$

$$1 \text{ yd}^3 = 27 \text{ ft}^3$$

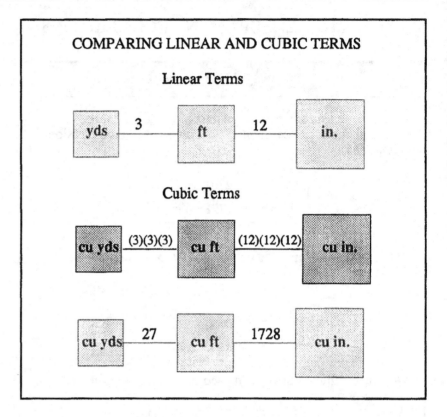

COMPARING LINEAR AND CUBIC TERMS

Linear Terms

Cubic Terms

Example 1: (Volume Measurement Conversions)
❑ The required volume for a screening pit is 325 cu ft. What is this volume in cubic yards?

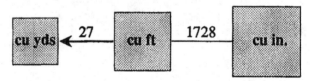

When converting from cu ft to cu yds we are moving from a larger box to a smaller box. Therefore, division is indicated:

$$\frac{325 \text{ cu ft}}{27 \dfrac{\text{cu ft}}{\text{cu yd}}} = \boxed{12 \text{ cu yds}}$$

---

* For a review of exponents, refer to Chapter 13, Powers, Roots, and Scientific Notation.

Example 2: (Volume Measurement Conversions)
❏ The capacity of a small segment of pipeline has been calculated to be 2512 cu in. How many cu ft is this?

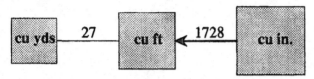

From cu in. to cu ft, we are moving to a smaller box. Therefore, division is indicated:

$$\frac{2512 \text{ cu in.}}{1728 \underset{\text{cu ft}}{\text{cu in.}}} = \boxed{1.5 \text{ cu ft}}$$

Example 3: (Volume Measurement Conversions)
❏ The volume of a trickling filter is 20,000 cu ft. To calculate the organic loading as gpd/ac-ft, the volume must be expressed as ac-ft. What is the ac-ft volume of the trickling filter?

From cu ft to ac-ft, we are moving to a smaller box. Thus, division is indicated:

$$\frac{20,000 \text{ cu ft}}{43,560 \underset{\text{cu yd}}{\text{cu ft}}} = \boxed{0.5 \text{ ac-ft}}$$

ACRES AND CU FEET CONVERSIONS

Volume is sometimes expressed as acre-feet. For example, many reservoir and waste treatment pond calculations include volume expressed as acre-feet.

The conversion equation to be used when converting between acre-feet and cubic feet is:

$$\boxed{1 \text{ ac-ft} = 43,560 \text{ cu ft}}$$

Notice that the same number (43,560) is used to convert between acres and square feet. This is because an acre-foot is one acre (or 43,560 sq ft) of water one foot deep:

1 ac-ft = (43,560 sq ft) (1 ft)

1 ac-ft = 43,560 cu ft

The box diagram that may be used in these calculations is:

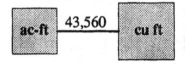

PRACTICE PROBLEMS 8.6: Volume Measurement Conversions

❏ Complete the following conversions.

1. 25 cu yds =_____cu ft

5. 92,600 cu ft =_____ac-ft

2. 1500 cu in. =_____cu ft

6. 17,260 cu ft =_____cu yds

3. 2.2 ac-ft =_____cu ft

7. 0.6 cu yds =_____cu ft

4. 21 cu ft =_____cu yds

8. 3 cu ft =_____cu in.

9. A screenings pit must have a capacity of 400 cu ft. How many cu yds is this?

ANS_____

10. A reservoir contains 50 ac-ft of water. How many cu ft of water does it contain?

ANS_____

8.7 mg/*L* AND PERCENT CONVERSIONS*

SUMMARY

Milligram per liter concentration may be expressed in terms of percent using the following equation:

$$1\% = 10,000 \text{ mg/}L$$

This equation may be written using the box method:

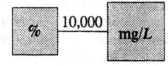

---

* mg/*L* to lbs/day calculations have been discussed in an entire chapter in the applied math texts (Chapter 3).

Milligrams per liter (mg/L) and percent (%) are two different ways of expressing concentration. In other words, the same concentration may be expressed in terms of mg/L or percent. If you know a concentration as a percent, you can also determine the concentration in mg/L. Or, if you know a concentration in mg/L, you can calculate the concentration in percent.

The relationship between milligrams per liter and percent is shown in the box to the right. In order to convert **from mg/L to percent, division by 10,000 is required**. The reverse is also true—to convert **from percent to milligrams per liter, multiplication by 10,000 is required**.

TWO EXPRESSIONS OF CONCENTRATION**

1. mg/L* is "parts per million" concentration

2. Percent* is "parts per hundred" concentration.

For example:

$$2500 \text{ mg} = \frac{2500. \text{ mg}}{1{,}000{,}000. \text{ mg}} = \frac{0.25 \text{ mg}}{100 \text{ mg}} = \boxed{0.25\%}$$

*In order to have 100 in the denominator—(for percent), the decimal point in the numerator and denominator must be moved **4 places to the left**. This is the same as **dividing each number by 10,000**.*

THE CONVERSION SHORTCUT

Using the following equation, percent and mg/L conversions may be completed quickly:

$$\boxed{1\% = 10{,}000 \text{ mg/L}}$$

Expressed as a box diagram, this is:

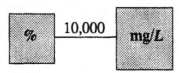

---

* For a review of milligrams per liter calculations, refer to Chapter 3 in the applied math texts. For a review of percent calculations, refer to Chapter 5.

** Grains/gal is also a measure of concentration and is discussed in Section 8.8.

Example 1: (mg/*L* and Percent Conversions)
❑ The suspended solids concentration of a primary clarifier is 340 mg/*L*. What is this concentration expressed as a percent?

Use the equation 1% = 10,000 mg/*L* to establish the box diagram:

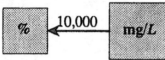

Since we are moving from a larger box to a smaller box, division by 10,000 is indicated:

$$\frac{340 \text{ mg}/L}{10,000 \text{ mg}/L/\%} = \boxed{0.034\%}$$

Example 2: (mg/*L* and Percent Conversions)
❑ A waste activated sludge has a total solids concentration of 0.6%. What is this expressed as mg/*L*?

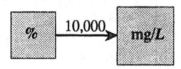

Moving from a smaller box to a larger box, multiplication by 10,000 is indicated:

$$(0.6\%)\,(10,000 \text{ mg}/L/\%) = \boxed{6000 \text{ mg}/L}$$

PRACTICE PROBLEMS 8.7: mg/*L* and Percent Conversions

❑ Complete the conversion problems given below.

1. 120 mg/*L* =_____%

3. 0.025% =_____mg/*L*

2. 1.5% =_____mg/*L*

4. 5000 mg/*L* =_____%

❑ Complete the following word problems.

5. The suspended solids concentration of the return activated sludge is 6800 mg/*L*. What is this concentration expressed as a percent?

ANS_____

6. A concentration of 195 mg/*L* is equivalent to a concentration of what percent?

ANS_____

8.8 mg/*L* AND gpg CONVERSIONS

SUMMARY

To convert from mg/*L* to grains per gallon* (gpg) or vice versa, the following equation is used:

$$1 \text{ gpg} = 17.1 \text{ mg/}L$$

This equation can be expressed as a box diagram:

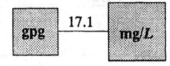

---

* US gallon

In the previous section, conversions for mg/*L* and percents were discussed. In addition to these two expressions of concentration, there is another expression used less frequently—grains per gallon (gpg).

However, you may be required to convert gpg concentration to mg/*L* concentration. The following conversion equation is used:

$$1 \text{ gpg} = 17.1 \text{ mg/}L$$

As with other conversion calculations, this relationship can be expressed as a box diagram:

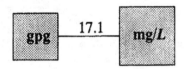

Example 1: (mg/*L* and gpg Conversions)
❑ If 13 gpg suspended solids enter the plant daily. How many mg/*L* is this?

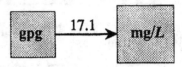

From gpg to mg/*L*, we are moving from a smaller box to a larger box; therefore multiplication by 17.1 is indicated:

$$\frac{(13 \text{ gpg}) (17.1 \text{ mg/}L)}{\text{gpg}} = \boxed{222.3 \text{ mg/}L}$$

Example 2: (mg/*L* and gpg Conversions)
❑ A concentration of 10 gpg is equivalent to how many mg/*L*?

Regardless of the direction of conversion, first draw the box diagram:

From gpg to mg/L box diagram with 17.1

Again, the conversion is from a smaller to larger box, and multiplication is indicated:

$$\frac{(10 \text{ gpg}) (17.1 \text{ mg/}L)}{\text{gpg}} = \boxed{171 \text{ mg/}L}$$

---

* For a review of milligrams per liter calculations, refer to Chapter 3 in the applied math texts.
  For a review of percent calculations, refer to Chapter 5.
** Grains/gal is also a measure of concentration and is discussed in Section 8.7.

Example 3: (mg/L and gpg Conversions)

❑ The trickling filter at a treatment plant removes 95 mg/L suspended solids. What is this concentration expressed as gpg?

Converting from mg/L to gpg, division by 17.1 is indicated:

$$\frac{95 \text{ mg/L}}{17.1 \frac{\text{mg/L}}{\text{gpg}}} = \boxed{5.6 \text{ gpg}}$$

Example 4: (mg/L and gpg Conversions)

❑ A concentration of 12 mg/L is equal to how many grains per gallon?

Division by 17.1 is indicated in this conversion problem:

$$\frac{12 \text{ mg/L}}{17.1 \frac{\text{mg/L}}{\text{gpg}}} = \boxed{0.7 \text{ gpg}}$$

PRACTICE PROBLEMS 8.8: mg/*L* and gpd Conversions

❏ Complete the conversion calculations indicated below.

1. 6.2 gpg =_____mg/*L*

3. 270 mg/*L* =_____gpg

2. 145 mg/*L* =_____gpg

4. 0.13 gpg =_____mg/*L*

❏ Complete the following word problems.

5. The suspended solids in a water is 2 gpg. What is this concentration in mg/*L*?

ANS_____

6. A suspended solids removal is 110 mg/*L*. What is this concentration in grains per gallon?

ANS_____

8.9 METRIC SYSTEM CONVERSIONS (METRIC TO METRIC)

SUMMARY

1. The metric prefixes to be used in all metric system conversions are:

Primary
Unit

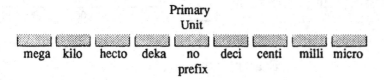

mega   kilo   hecto   deka   no   deci   centi   milli   micro
prefix

2. For conversions involving units of grams, liters or meters,* **move the decimal point one place** to the right or left **for each place value** move to the right or left, until you have reached the position of the desired unit.

3. For conversions involving units in square terms, **move the decimal point two places** to the right or left **for each place value** move to the right or left, until you have reached the position of the desired unit.

4. For conversions involving units in cubic terms, **move the decimal point three places** to the right or left **for each place value** move to the right or left, until you have reached the position of the desired unit.

---

* For linear measurements, only square and cubic terms require a different decimal point move, as listed above in 3 and 4.

Conversions from one unit to another within the metric system are very simple. These conversions do require, however, that you are familiar with metric system prefixes and their position in the place value system, shown to the right. It is advisable that you memorize this information.

The metric system is based on units or "powers" of ten.* Therefore, converting from one unit to another simply involves moving the decimal point to the right or left, according to the place value move. Note: **no calculation is required**.

To convert from one unit to another:

1. Locate the place value of the units you wish to convert.

2. Locate the place value of the desired unit.

3. Move the decimal point to the right or left the same number of places as indicated in the place value system diagram.

Hint: When no decimal point is shown for a number, such as 202 or 60, the decimal point is assumed to be at the right of the numbers shown, such as 202.0 or 60.0.

METRIC SYSTEM PREFIXES AND VALUES

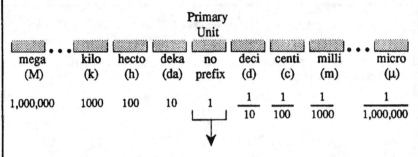

mega (M)		kilo (k)	hecto (h)	deka (da)	Primary Unit no prefix	deci (d)	centi (c)	milli (m)		micro (μ)
1,000,000		1000	100	10	1	$\frac{1}{10}$	$\frac{1}{100}$	$\frac{1}{1000}$		$\frac{1}{1,000,000}$

Examples include:

meters—linear measurement
liters—capacity measurement
grams—weight measurement

EACH PLACE VALUE MOVE REQUIRES ONE DECIMAL POINT MOVE

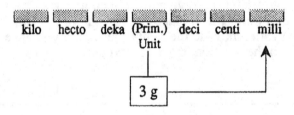

| kilo | hecto | deka | (Prim.) Unit | deci | centi | milli |

3 g

For example, suppose you wish to convert 3 grams to milligrams. Grams and milligrams are separated by **three place values to the right**. Therefore, the decimal point must be moved three places to the right:

3.000 = 3000 mg
1 2 3

3 g = 3000 mg

---

* For a review of powers, refer to Chapter 13.

Example 1: (Metric System Conversions)
❑ Convert 2500 milliliters to liters.

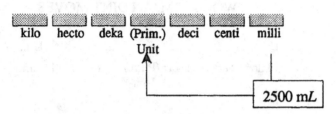

Converting from milliliters to liters, requires a move of **three place values to the left**. Therefore, the decimal point must be moved three places to the left:

$$2500. = \boxed{2.5 \text{ liters}}$$

Example 2: (Metric System Conversions)
❑ Convert 1.8 kilograms to grams.

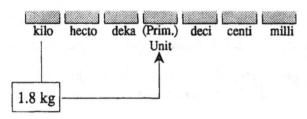

From kilograms to grams, there is a move of **three place values to the right**. The decimal point is therefore moved three places to the right:

$$1.800 = \boxed{1800 \text{ grams}}$$

PREFIXES AND ABBREVIATIONS

The basic measurement units (the primary units) of the metric system are:

- meters (m)
  (for measurement of length)

- liters (*L*)
  (for measurement of capacity or volume)

- grams (g)
  (for measurement of weight)

Prefixes are added to indicate multiples or fractions of the basic unit.
Each primary unit has eight prefixes which could be used with that unit, as follows:

Prefix	Abbreviation When Used With Primary Unit		
	Meter (m)	Liter (*L*)	Gram (g)
mega (M)	Mm	M*L*	Mg
kilo (k)	km	k*L*	kg
hecto (h)	hm	h*L*	hg
deka (da)	dam	da*L*	dag
deci (d)	dm	d*L*	dg
centi (c)	cm	c*L*	cg
milli (m)	mm	m*L*	mg
micro (μ)	μm	μ*L*	μg

For square or cubic terms (**used only with the primary unit of meters**), the word "square" or "cubic" is used with the appropriate prefix, or an exponent is used (for example square kilometers or km^2).

CONVERTING SQUARE TERMS

When converting square terms, such as square meters (m^2) or square kilometers (km^2), the decimal move is a little different than that described on the preceding pages.

Instead of moving one decimal place for each place value move, **the decimal point must be moved two places for each place value move.**

The use of exponents* in abbreviating terms (such as m^2, km^2, etc) can help you remember to move in **groups of two.** The examples on this page illustrate conversions in square terms.

FOR SQUARE TERMS EACH PLACE VALUE MOVE REQUIRES TWO DECIMAL POINT MOVES

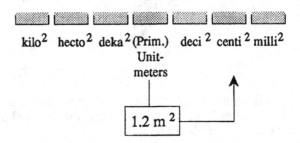

$$kilo^2 \quad hecto^2 \quad deka^2 \quad \begin{array}{c}(Prim.)\\ Unit-\\ meters\end{array} \quad deci^2 \quad centi^2 \quad milli^2$$

$$\boxed{1.2 \text{ m}^2}$$

Suppose you want to convert 1.2 square meters to square centimeters. From meters to centimeters there are two place value moves. But for each place value move in square terms, the decimal point must be moved two places. The easiest way to keep this straight is to move in **groups of two:**

$$1.2000 = \boxed{12,000 \text{ cm}^2}$$

Example 3: (Metric System Conversions)
❑ Convert 130,000 square meters to square kilometers.

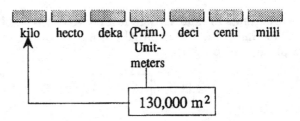

$$kilo \quad hecto \quad deka \quad \begin{array}{c}(Prim.)\\ Unit-\\ meters\end{array} \quad deci \quad centi \quad milli$$

$$\boxed{130,000 \text{ m}^2}$$

There are three place value moves to the left. Therefore the decimal point must be moved three **groups of two** to the left:

$$130,000. = \boxed{0.13 \text{ km}^2}$$

---

* Refer to Chapter 13 for a review of exponents.

CONVERTING CUBIC TERMS

The decimal point moves are also different when converting cubic terms, such as cubic meters (m^3) or cubic centimeters (cm^3).* **For each place value move, the decimal point must be moved three places.**

The use of exponents in abbreviating terms (such as m^3, km^3, etc.) can help you remember to move in **groups of three.**

FOR CUBIC TERMS EACH PLACE VALUE MOVE REQUIRES THREE DECIMAL POINT MOVES

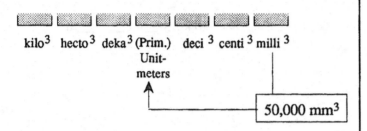

$kilo^3$ $hecto^3$ $deka^3$ (Prim.) $deci^3$ $centi^3$ $milli^3$
Unit-
meters

50,000 mm³

For example, from millimeters to meters, there are three place value moves to the left. Because this conversion is in cubic terms, for each of these place value moves, the decimal point must be moved three places. Move the decimal point in **groups of three:**

$$.000050000. = \boxed{0.00005 \text{ m}^3}$$

Example 4: (Metric System Conversions)
❑ Convert 1 cubic meter to cubic centimeters.

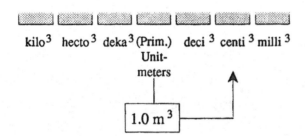

$kilo^3$ $hecto^3$ $deka^3$ (Prim.) $deci^3$ $centi^3$ $milli^3$
Unit-
meters

1.0 m³

There are two place value moves to the right. The decimal point must therefore be moved **two groups of three** to the right:

$$1.000000. = \boxed{1,000,000 \text{ cm}^3}$$

---

* In the medical field, cubic centimeters is abbreviated as "cc".

PRACTICE PROBLEMS 8.9: Metric System Conversions (Metric to Metric)

❑ Complete the metric system conversion calculations indicated below.

1. 2400 mL =_____L

2. 8 g =_____kg

3. 15 L =_____mL

4. 2.2 mL =_____L

5. 180 mm =_____m

6. 40 km =_____m

7. 230 mg =_____g

8. 250,000 m^2 =_____km^2

9. 155,000 mm^3 =_____cm^3

10. 30 kg =_____g

8.10 METRIC/ENGLISH CONVERSIONS

SUMMARY

To convert from the metric to English system, or vice versa:

1. Find the conversion equation that relates the units given to those desired. (If no conversion equation is given for precisely those two units, be sure that **at least the English units are in the correct terms**. It is much easier to make any final adjustments in metric units rather than English units.)

2. Use the box method of conversions* to complete the English to metric or metric to English conversion.

3. Make final metric system conversions, as needed.

---

* For a description of this method, refer to Section 8.1.

Conversions between the metric and English systems are similar to conversions described in Sections 8.2 through 8.8. As with other conversions, the "box method of conversions"* can be used.

In making a conversion, first select the conversion equation that includes both the given and desired units. Then use the box method to calculate the conversion.

Example 1: (Metric/English Conversions)
❑ 20 ft is equivalent to how many meters?

From the table below, there are two equations that include both feet and meters. The desired equation, however, is the one in which *the numbers on both sides of the equation are equal to or greater than one*: 1 m = 3.281 ft.

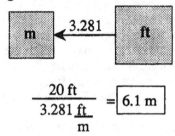

$$\frac{20 \text{ ft}}{3.281 \frac{\text{ft}}{\text{m}}} = \boxed{6.1 \text{ m}}$$

CONVERSION EQUATIONS

Metric to English		**English to Metric**	
LINEAR MEASURE			
1 centimeter (cm) = 0.3937 inches		1 inch = 2.540 cm	
1 meter (m) = 3.281 feet		1 foot = 0.3048 m	
1 meter (m) = 1.0936 yards		1 yard = 0.9144 m	
1 kilometer (km) = 0.6214 miles		1 mile = 1.609 km	
SQUARE MEASURE			
1 sq centimeter = 0.155 sq in		1 sq in = 6.4516 sq cm	
1 sq meter = 10.76 sq ft		1 sq ft = 0.0929 sq m	
1 sq meter = 1.196 sq yd		1 sq yd = 0.8361 sq m	
CUBIC MEASURE			
1 cu centimeter = 0.061 cu in		1 cu in = 16.39 cu cm	
1 cu meter = 35.3 cu ft		1 cu ft = 0.0283 cu m	
1 cu meter = 1.308 cu yd		1 cu yd = 0.7645 cu m	
CAPACITY			
1 liter (*L*) = 61.025 cu in		1 cu in = 0.0164 *L*	
1 liter (*L*) = 0.0353 cu ft		1 cu ft = 28.32 *L*	
1 liter (*L*) = 0.2642 gal (US)		1 gal (US) = 3.785 *L*	
WEIGHT			
1 gram (g) = 15.43 grains (gr)		1 grain = 0.0648 g	
1 gram (g) = 0.0353 ounces		1 ounce = 28.35 g	
1 kilogram (kg) = 2.205 pounds		1 pound = 453.6 g	
		1 pound = 7000 grains	

---

* For a discussion of the "box method of conversions", refer to Section 8.1.

Example 2: (Metric/English Conversions)
❑ Express 50 liters in terms of gallons.

First, select the equation that includes both liters and gallons. There are two possible options: $1\,L = 0.2642$ gal and 1 gal $= 3.785\,L$.

Always select the equation that has numbers greater than or equal to one on both sides of the equation:

$$1 \text{ gal} = 3.785\,L$$

$$\frac{50\,L}{3.785\,\dfrac{L}{\text{gal}}} = \boxed{13.2 \text{ gal}}$$

Example 3: (Metric/English Conversions)
❑ Express 5.6 feet in terms of centimeters.

There is no equation in the table that relates feet and centimeters. Select the equation that has the English units (feet) in correct terms. Be sure that both numbers in the equation are greater than or equal to one:

$$1 \text{ m} = 3.281 \text{ ft}$$

$$\frac{5.6 \text{ ft}}{3.281\,\dfrac{\text{ft}}{\text{m}}} = \boxed{1.7 \text{ m}}$$

The metric unit (m) can now be easily converted to centimeters:

(Prim.) deci centi
Unit-

$$\boxed{1.7 \text{ m}}$$

Move the decimal point two places to the right:

$$1.70 = \boxed{170 \text{ cm}}$$

The conversion equations given in the table include both metric to English and English to metric. Using the box method of conversions, **it is important to select the equation that has numbers greater than one on both sides of the equation.**

Occasionally, the only equation that relates the given and desired units will include a number less than or equal to one (a decimal fraction). Since such an equation will not work with the box method of conversions, the equation will have to be modified. Simply divide both sides of the equation by the decimal fraction. Using the equation from Example 2 as an example:

$$\boxed{1\,L = 0.2642 \text{ gal}}$$

Divide both sides of the equation by 0.2642:

$$\frac{1\,L}{0.2642} = \frac{0.2642 \text{ gal}}{0.2642}$$

$$3.785\,L = 1 \text{ gal}$$

This can be rewritten with the one on the left side of the equation:

$$\boxed{1 \text{ gal} = 3.785\,L}$$

In some cases you will not be able to find a conversion equation that includes both the given and desired units. When this occurs, **be sure to select the equation that includes the correct English unit.** Conversions are much easier in the metric system than the English system. Example 3 illustrates a conversion calculation where the conversion equation does not match the given and desired terms.

PRACTICE PROBLEMS 8.10: Metric/English Conversions

❑ Complete the following metric/English conversions, using the conversion equation chart given on the previous page.

1. 70 cm = _____ in.

2. 40 cu ft = _____ m³

3. 35 yds = _____ m

4. 200 m² = _____ sq ft

5. 25 g = _____ oz

6. 6200 lbs = _____ kg

7. 20 oz = _____ g

8. 600 mL = _____ gal

9 *Linear Measurement*

Complete and score the following skills test. Each section should be scored separately in the box provided to the right. A score of 4 or above indicates you are sufficiently strong in that concept. A score of 3 or below indicates a review of that section is advisable.

9.1 Perimeter Calculations

Number
Correct

❑ What is the perimeter of the areas shown below?

1.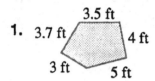

ANS_____

2.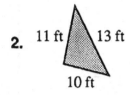

ANS_____

3.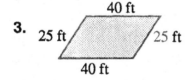

ANS_____

❑ Calculate the length of the unknown side, given the following information:

4.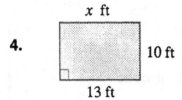

ANS_____

5. Three sides of an object measure 25 ft, 78 ft, and 55 ft. What is the length of the fourth side if the perimeter measures 240 ft?

ANS_____

9.2 Circumference Calculations

Number Correct

❑ What is the circumference of the circles given below?

1.

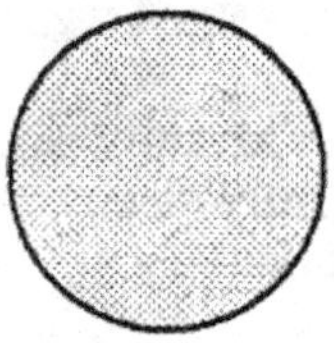

Diameter = 20 ft

ANS________

2.

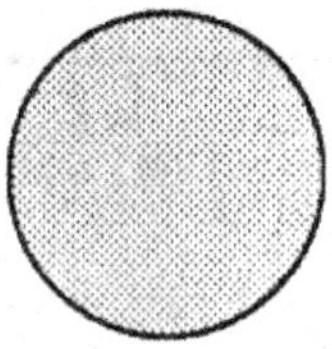

Diameter = 85 ft

ANS________

3.

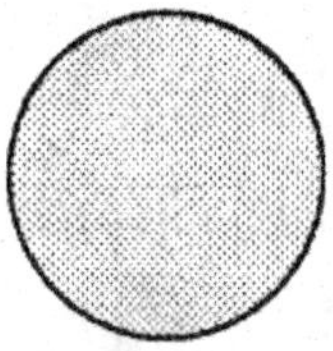

Radius = 30 ft

ANS________

4. Calculate the diameter of the circle shown below.

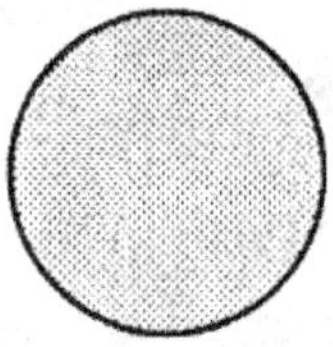

Circumference = 157 ft

ANS________

5. If the circumference of a circle is 251.2 ft, what is the radius of that circle?

ANS________

9.1 PERIMETER CALCULATIONS

SUMMARY

1. To determine the perimeter (the distance around) of any angular area or object, add the length of each side. This is the **general perimeter equation.**

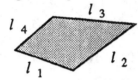

Perimeter = Length + Length + Length + Length + ...
               Side 1   Side 2   Side 3   Side 4

$$P = l_1 + l_2 + l_3 + l_4 + ...$$

2. To determine the perimeter of a square, you may use the general equation above, or the modified equation listed below:

$$P = 4s$$

3. To determine the perimeter of a rectangle, you may use the general equation or the modified equation as follows:

$$P = 2l + 2w$$

Linear measurement is simply the measurement along a line. These lengths may be expressed using the English System of measurement, such as inches, feet, yards, and miles, or using the Metric System of measurement, such as millimeters, centimeters, meters, and kilometers.*

Many water and wastewater calculations require tank or channel dimensions, pipe lengths and diameters, weir lengths, and other linear measurements. Although these dimensions are normally provided in the treatment system plans and specifications, it may be wise to verify the lengths indicated.

This chapter focuses on one particular type of linear measurement: the distance around the outside edge of an area or object—**the perimeter and circumference.**

---

* Calculations presented in this book will use English System units of measurement. For a discussion of calculations using Metric System measurements, refer to Chapter 8—Conversions.

ADD LENGTHS OF SIDES

The distance around an angular object or area is called the perimeter. **To calculate the perimeter** of any area or object **add the length of each of its sides:**

Perimeter = $l_1 + l_2 + l_3 + ...$

The number of terms added for the perimeter equation depends on how many sides the object has. For example, if an object has three sides, the lengths of three sides are added, if it has six sides, then six lengths must be added to determine the perimeter.

FINDING THE LENGTH OF ONE SIDE

The perimeter equation can be used to determine the length of one of the sides. To determine the length of the unknown side subtract the total of all other sides from the perimeter length. For example, if the distance around an object with four sides is 20 ft, and the length of three sides totals 16 ft, the length of the fourth side must be

20 ft – 16 ft = 4 ft.

Another calculation of this type is given in Example 2.

Example 1: (Perimeter)

❑ What is the perimeter of the figure shown below?

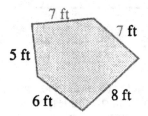

$$P = l_1 + l_2 + l_3 + l_4 + l_5$$

$$= 5 \text{ ft} + 6 \text{ ft} + 8 \text{ ft} + 7 \text{ ft} + 7 \text{ ft}$$

$$= \boxed{33 \text{ ft}}$$

Example 2: (Perimeter)

❑ The perimeter of the figure shown below is 56 ft. What is the length of the fifth side if four of the sides have lengths as shown?

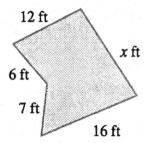

Use the same general perimeter equation for this type of problem:

$$P = l_1 + l_2 + l_3 + l_4 + l_5$$

$$56 \text{ ft} = 12 \text{ ft} + 6 \text{ ft} + 7 \text{ ft} + 16 \text{ ft} + x \text{ ft}$$

$$56 \text{ ft} = 41 \text{ ft} + x \text{ ft}$$

$$56 \text{ ft} - 41 \text{ ft} = x \text{ ft}$$

$$\boxed{15 \text{ ft}} = x \text{ ft}$$

Example 3: (Perimeter)
❏ Calculate the perimeter of the square shown below.

$$s = 22 \text{ ft}$$

$$P = 4s$$
$$= (4)\ (22 \text{ ft})$$
$$= 88 \text{ ft}$$

Note that the general equation for perimeter would also have resulted in the same answer, although it is generally a slower process:

$$P = l_1 + l_2 + l_3 + l_4$$
$$= 22 \text{ ft} + 22 \text{ ft} + 22 \text{ ft} + 22 \text{ ft}$$
$$= \boxed{88 \text{ ft}}$$

PERIMETER OF SQUARES

Certain shapes, such as squares and rectangles have slightly modified equations for perimeter due to some distinctive aspect of their shapes. For example, **a square always has four equal sides.** Although the general perimeter equation would be adequate ($P = l_1 + l_2 + l_3 + l_4$), it does not take advantage of the fact that all the lengths are equal. An equation often used for the perimeter of a square is:

$$P = 4s$$

The "s" is used to denote "side length," although an "l" could be used, if desired ($P = 4l$). Once you know the length of only one side of a square, you can determine its perimeter.

Example 4: (Perimeter)
❏ What is the perimeter of the rectangle shown below?

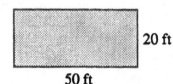

20 ft

50 ft

$$P = 2l + 2w$$
$$P = (2)\ (50 \text{ ft}) + (2)\ (20 \text{ ft})$$
$$= 100 \text{ ft} + 40 \text{ ft}$$
$$= \boxed{140 \text{ ft}}$$

Using the general perimeter equation, you would first have to determine the other two sides (one would be 50 ft and the other 20 ft). The calculation would be:

$$P = l_1 + l_2 + l_3 + l_4$$
$$= 50 \text{ ft} + 20 \text{ ft} + 50 \text{ ft} + 20 \text{ ft}$$
$$= \boxed{140 \text{ ft}}$$

PERIMETER OF RECTANGLES

The perimeter equation for a rectangle is also a special case of the general perimeter equation. Since the two lengths and two widths are equal, the equation often used to calculate the perimeter of a rectangle is:

$$P = 2l + 2w$$

As with other perimeter calculations, this equation can be used to calculate a perimeter, or it can be used to calculate the length of one of the sides (the total length of the perimeter would have to be known).

These special equations are particularly useful in solving problems where the length of one of the sides is unknown.

PRACTICE PROBLEMS 9.1: Perimeter Calculations

❑ Find the perimeter for each figure given below.

1.

8 ft   9 ft

6 ft

ANS_____

2.

75 ft

50 ft

55 ft   100 ft

90 ft

ANS_____

3.

30 ft

(A square)

ANS_____

4.

7 ft

9 ft

10 ft

7 ft

8 ft

ANS_____

5.

70 ft

20 ft

ANS_____

(The marks in the corners indicate right angles; therefore, the figure is a rectangle)

6. The lengths of each side of a fenced area are as follows: 87 ft, 100 ft, 82 ft, and 105 ft. What is the perimeter of the fenced area?

ANS_____

7. The length of a rectangle is 8 inches and the width is 5 inches. What is the perimeter of the rectangle?

ANS_____

8. Three sides of an object have lengths of 20 in., 82 in., and 25 in. If the perimeter of the object measures 215 in, what is the length of the fourth side?

ANS_____

9. The length of one side of an aeration basin is 150 ft. If the perimeter is 350 ft, what is the length of the other three sides?

ANS_____

10. If one side of a square measures 6 inches, its perimeter would be how many inches?

ANS_____

9.2 CIRCUMFERENCE CALCULATIONS

In the previous section, we calculated the distance around angular areas by adding the lengths of all the sides. However, since circles do not have side lengths, a different calculation is required to determine the distance around the circle. This distance, called the circumference of the circle, is calculated using the equation given to the left.

The most common calculation of circumference in water and wastewater math is perhaps the calculation of weir circumference, that is necessary for calculating the weir overflow rate.

SUMMARY

1. To determine the circumference of (distance around) a circular area or object, multiply the diameter by Pi (π), 3.14:

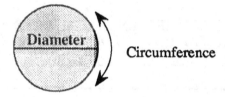

$$\text{Circumference} = (\pi) \ (\text{Diameter})$$

$$= (3.14) \ (\text{Diameter})$$

2. For a quick estimate of the circumference, multiply the diameter by three. For example, if the diameter of a tank is 20 ft, the distance around the tank (circumference) is about 20 ft x 3 $\approx$ 60 ft (a "wavy" equals sign is used to denote "approximately equal to").

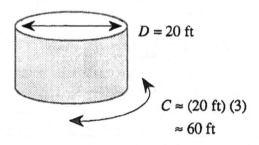

$$D = 20 \text{ ft}$$

$$C \approx (20 \text{ ft}) \ (3)$$

$$\approx 60 \text{ ft}$$

CIRCUMFERENCE AND DIAMETER

One of the best ways to understand the equation for the circumference of a circle is to understand the relationship of the diameter of a circle to its circumference.

The diagram to the right illustrates that as the diameter gets larger, so does the circumference. In fact, in each case shown, the circumference is about three times the length of the diameter. For all circles, the distance around the circle (circumference) is about three times the distance across the circle (diameter).

QUICK ESTIMATE OF THE CIRCUMFERENCE

For a quick estimate of the circumference of a circle, simply multiply the diameter by three.

AS THE DIAMETER INCREASES THE CIRCUMFERENCE INCREASES

  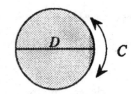

$D = 2$ ft $D = 5$ ft $D = 20$ ft
$C = 6.28$ ft $C = 15.7$ ft $C = 62.8$ ft

THE CIRCUMFERENCE IS ALWAYS ABOUT THREE TIMES THE LENGTH OF THE DIAMETER

Example 1: (Circumference)
❑ The diameter of a circle is 50 ft. What is the approximate distance around the circle?

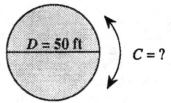

In every circle, the circumference is about 3 times the diameter:

$$(50 \text{ ft}) (3) = \boxed{150 \text{ ft}}$$

approximate
circumference

Example 2: (Circumference)
❑ If the radius of a circle is 5 ft, what is the approximate circumference of that circle?

Radius = 5 ft
Therefore:
Diameter = 10 ft

$$(10 \text{ ft}) (3) = \boxed{30 \text{ ft}}$$

approximate
circumference

Example 3: (Circumference)

❑ The diameter of a tank is 60 ft. What is the distance around the tank?

$D = 60$ ft

Circumference $= \pi D$

$= (3.14)(60 \text{ ft})$

$= \boxed{188.4 \text{ ft}}$

Example 4: (Circumference)

❑ The radius of a tank is 35 ft. What is the circumference of the tank?

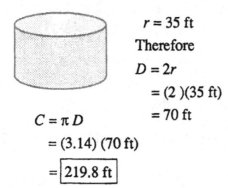

$r = 35$ ft

Therefore

$D = 2r$

$= (2)(35 \text{ ft})$

$= 70$ ft

$C = \pi D$

$= (3.14)(70 \text{ ft})$

$= \boxed{219.8 \text{ ft}}$

Example 5: (Circumference)

❑ The circumference of a tank is known to be 266.9 ft. What is the length of the diameter of that tank?

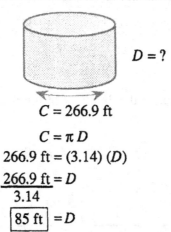

$D = ?$

$C = 266.9$ ft

$C = \pi D$

$266.9 \text{ ft} = (3.14)(D)$

$\dfrac{266.9 \text{ ft}}{3.14} = D$

$\boxed{85 \text{ ft}} = D$

CALCULATION OF THE CIRCUMFERENCE

The circumference of any circle is about three times the length of its diameter. More precisely, it is about 3.14 times its diameter. This constant ratio between the lengths of the circumference and diameter of any circle is called Pi, represented by the Greek letter π. The equation for calculating the circumference can be written as:

$C = \pi D$, or

$C = (3.14)(D)$

THE RADIUS IS HALF A DIAMETER

Occasionally the distance across a circular area or object will be given in terms of the radius rather than the diameter, as shown in Example 2 and 4. A radius is the distance from the center of the circle to the outside edge. It is always one-half the length of the diameter. Convert the radius to diameter length and continue with the calculation.

CALCULATING THE DIAMETER

If you know the circumference of a circle, you can calculate its diameter using the same equation,

$C = \pi D$,

as shown in Example 5. (*Note*: this problem involves solving for the unknown value. If you need further explanation of this process, refer to Chapter 2)

PRACTICE PROBLEMS 9.2: Circumference Calculations

❏ Using the "quick estimate" method to determine circumference, what is the approximate circumference of the circles shown below?

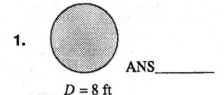

1.   ANS_____

$D = 8$ ft

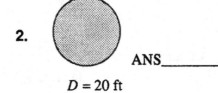

2.   ANS_____

$D = 20$ ft

❏ Calculate the circumference for the circles shown below.

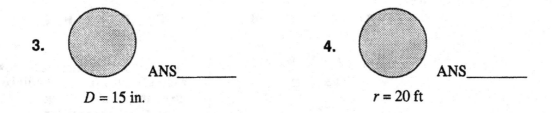

3.   ANS_____

$D = 15$ in.

4.   ANS_____

$r = 20$ ft

5. The weir diameter of a clarifier is 50 ft. What is the total ft of weir for the clarifier?

ANS_____

6. The radius of a tank is 2 ft. What is the circumference of the tank?

ANS_____

7. If the circumference of a tank is 376.8 ft, what is the diameter of the tank?

ANS_____

10   *Area Measurement*

Complete and score the following skills test. Each section should be scored separately in the box provided to the right. For Section 10.1, a score of 8 or above indicates you are sufficiently strong in that concept. A score of 7 or below indicates a review of that section is advisable. For Sections 10.2 and 10.3, a score of 4 or above indicates you are sufficiently strong in that concept. A score of 3 or below indicates a review of that section is advisable.

10.1   Areas—Basic Shapes

Number
Correct

❏ Calculate the sq ft areas for the figures shown below.

1.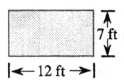

ANS_____

4.

Diam = 8 ft

ANS_____

2.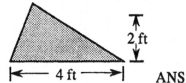

ANS_____

5.

(square)

ANS_____

3.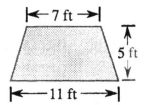

ANS_____

6.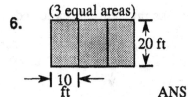

ANS_____

❏ Solve the following problems, as indicated.

7. A clarifier is 75 ft long and has a surface area of 1875 sq ft. What is the width of the clarifier?

ANS_____

8. What is the sq in. cross section of an 8-inch pipe?

ANS_____

9. The cross section of a trapezoidal channel is shown below. Given the dimensions as shown, what is the distance across the top of the water? (Area = 21 sq ft)

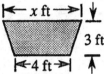

3 ft

|←4 ft→|

ANS_____

10. What is the cross-sectional area (sq in) of the trough shown below?

|←6 in→|

ANS_____

10.2 Areas—Combined Shapes

❑ Calculate the sq ft areas for the figures shown below.

1.

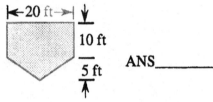

|←20 ft→|

10 ft

5 ft

ANS_____

3.

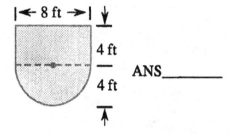

|← 8 ft →|

4 ft

4 ft

ANS_____

2.

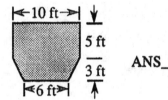

|←10 ft→|

5 ft

3 ft

|←6 ft→|

ANS_____

❑ Solve the following problems, as indicated.

4. What is the total depth of water in the channel, given the information shown below.
(Total Area = 8 sq ft)

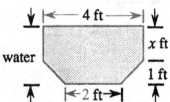

|← 4 ft →|

water

x ft

1 ft

|←2 ft→|

ANS_____

5. What is the length of *x* in the diagram shown below?

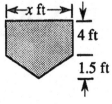

|←x ft→|

4 ft

Total Area = 38 sq ft

1.5 ft

ANS_____

10.3 Lateral Areas

❑ Calculate the lateral area for each figure shown below.

1. 3 ft

|← 4 ft →|

ANS_____

2. 10 ft

|← 35 ft →|

ANS_____

3. 5 ft

ANS_____

❑ Complete the following problems, as indicated.

4. The lateral area of a cylinder is 628 sq ft. If the height of the tank is 8 ft, what is the length of its diameter?

ANS_____

5. The diameter of a cone is 6 ft. If the lateral area of the cone is 47 sq ft, what is the slant height of the cone?

ANS_____

NOTES:

10.1 AREAS—BASIC SHAPES

SUMMARY

The equations for the four basic shapes most often used in area calculations are as follows:

Rectangle

$$A_\square = lw$$

Triangle

$$A_\triangle = \frac{bh}{2}$$

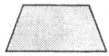

Trapezoid

$$A_\triangle = \frac{(b_1 + b_2)h}{2}$$

Circle

$$A_\bigcirc = (0.785)(D^2)$$

Area measurement is the measurement of the amount of space on the SURFACE of an object. Since the square is the basis by which these measurements are made, the units used to express this surface space are square inches (sq in or in^2).

The four area equations shown to the left are used frequently in water and wastewater calculations and should be memorized. They are the basis of most volume calculations as well.

AREA OF A RECTANGLE

The area of a rectangle is calculated by multiplying the length of the rectangle by its width:

$$A_\square = lw$$

This calculation is used repeatedly in water and wastewater math. In addition, it is the basis for understanding the calculations of triangle and trapezoid area.

Example 1: (Area of Rectangle)
❏ A rectangle has a length of 5 inches and a width of 3 inches. What is the sq in area of the rectangle?

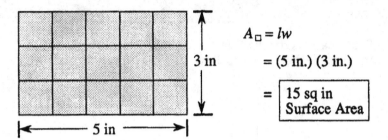

$$A_\square = lw$$
$$= (5 \text{ in.}) (3 \text{ in.})$$
$$= \boxed{\begin{array}{l} 15 \text{ sq in} \\ \text{Surface Area} \end{array}}$$

Example 2: (Area of Rectangle)
❏ What is the sq ft area of a rectangle 5 ft by 4 ft?

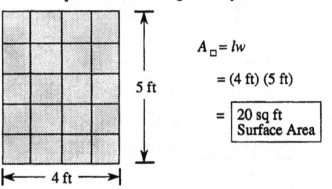

$$A_\square = lw$$
$$= (4 \text{ ft}) (5 \text{ ft})$$
$$= \boxed{\begin{array}{l} 20 \text{ sq ft} \\ \text{Surface Area} \end{array}}$$

Example 3: (Area of Rectangle)
❏ A lift station wet well is 10 feet long and 10 feet wide. What is the surface area of the wet well?

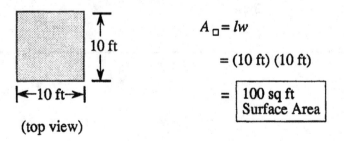

(top view)

$$A_\square = lw$$
$$= (10 \text{ ft}) (10 \text{ ft})$$
$$= \boxed{\begin{array}{l} 100 \text{ sq ft} \\ \text{Surface Area} \end{array}}$$

Example 4: (Area of Rectangle)

❑ The surface area of a tank is 2000 sq ft. If the width of the tank is 25 ft, what is the length of the tank?

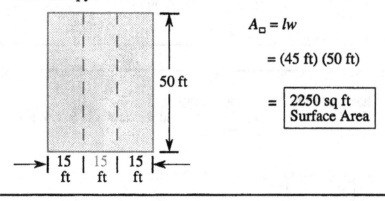

$$A_\square = lw$$

$$2000 \text{ sq ft} = (x)(25 \text{ ft})$$

$$\frac{2000 \text{ sq ft}}{25 \text{ ft}} = x$$

$$\boxed{\begin{array}{c} 80 \text{ ft} \\ \text{Length} \end{array}} = x$$

SOLVING FOR THE LENGTH OR WIDTH

The equation for the area of a rectangle can be used to calculate any of the three variables (A, l or w), provided information is given for the other two variables. Begin with the same equation, regardless of which term is unknown, fill in the known information, then solve for the unknown variable. (For a review of solving for the unknown term, refer to Chapter 2.)

Example 5: (Area of Rectangle)

❑ A treatment plant has three drying beds, each of which is 50 ft long and 15 ft wide. How many sq ft do the drying beds occupy?

50 ft

15 ft | 15 ft | 15 ft

$$A_\square = lw$$

$$= (45 \text{ ft})(50 \text{ ft})$$

$$= \boxed{\begin{array}{c} 2250 \text{ sq ft} \\ \text{Surface Area} \end{array}}$$

AREA OF A TRIANGLE

The diagram to the right illustrates the relationship of the areas of a rectangle and a triangle. The area of a triangle is always one-half the area of a rectangle with the same length and width dimensions.
Therefore, the equation for the area of a triangle can be stated as:

$$A_\triangle = \frac{lw}{2}$$

Often, however, the equation for the area of a triangle is written as:

$$A = \frac{bh}{2}$$

The *b* for **base** and *h* for **height** are preferred since it helps emphasize the fact that the **height of the triangle must be measured vertically from the horizontal base.**

EVERY TRIANGLE IS ONE-HALF A RECTANGLE

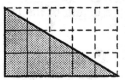

$$\text{Area of Rectangle} = lw$$
$$= (10 \text{ in}) (6 \text{ in})$$
$$= \boxed{60 \text{ sq in}}$$
$$\text{Area of Triangle} = \frac{lw}{2}$$
$$= \frac{(10 \text{ in}) (6 \text{ in})}{2}$$
$$= \boxed{30 \text{ sq in}}$$

Example 1: (Area of a Triangle)
❏ The base of a triangle is 4 ft and the height is 5 ft. What is the sq ft area of the triangle?

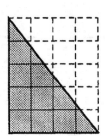

$$A = \frac{bh}{2}$$
$$= \frac{(4 \text{ ft}) (5 \text{ ft})}{2}$$
$$= \boxed{\begin{array}{l} 10 \text{ sq ft} \\ \text{Surface Area} \end{array}}$$

Example 2: (Area of Triangle)
❑ What is the sq ft area of a triangle that has a base of 6 ft and a height of 3 ft?

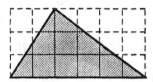

$$A = \frac{bh}{2}$$

$$= \frac{(6 \text{ ft}) (3 \text{ ft})}{2}$$

$$= \boxed{\begin{array}{l} 9 \text{ sq ft} \\ \text{Surface Area} \end{array}}$$

CALCULATING THE BASE OR HEIGHT

There are three variables in the equation for the area of a triangle: the area (A), the base (b), and the height (h). The same general equation can be used to calculate any one of the variables, as long as information is given for the other two variables. (If you need additional explanation in solving for the unknown term, refer to Chapter 2.)

Example 3: (Area of Triangle)
❑ A triangular portion of the treatment grounds is not being used. If the area is 17,000 sq ft and the base of the area is 200 ft, what is the height of the area?

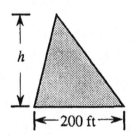

← 200 ft →

$$A = \frac{bh}{2}$$

$$17,000 \text{ sq ft} = \frac{(200 \text{ ft}) (x \text{ ft})}{2}$$

$$(17,000 \text{ sq ft}) (2) = (200 \text{ ft}) (x \text{ ft})$$

$$\frac{(17,000) (2)}{200} = x \text{ ft}$$

$$\boxed{170 \text{ ft}} = x \text{ ft}$$

AREA OF A TRAPEZOID

The trapezoid area is actually a variation of the rectangle area, and it might be termed an "averaged rectangle." The two length dimensions are added then divided by two to determine the **average length dimension.** In the problem shown to the right, the two lengths (12 ft and 8 ft) are added together then divided by two. The answer (10 ft) is the average length which can then be multiplied by the height of the trapezoid (10 ft x 6 ft = 60 sq ft).

A TRAPEZOID IS AN "AVERAGED RECTANGLE"

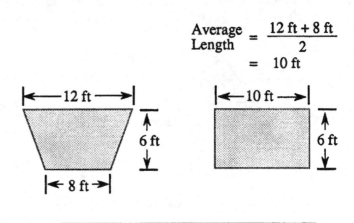

$$\text{Average Length} = \frac{12\ ft + 8\ ft}{2}$$

$$= 10\ ft$$

$$\boxed{\begin{array}{l} \text{Area of} \\ \text{Trapezoid} \end{array} = (\text{Average length})(\text{width})}$$

As with the triangle equation, the terms "base" and "height" are preferred over "length" and "width", respectively so that the equation becomes:

$$\boxed{\begin{array}{l} \text{Area of} \\ \text{Trapezoid} \end{array} = \frac{(b_1 + b_2)(\text{height})}{2}}$$

In the figure given above, the area calculation would be:

$$A = \frac{(8\ ft + 12\ ft)(6\ ft)}{2}$$

$$= \frac{(20\ ft)(6\ ft)}{2}$$

$$= (10\ ft)(6\ ft)$$

$$= \boxed{\begin{array}{l} 60\ sq\ ft \\ \text{Surface Area} \end{array}}$$

---

* For a review of the concept of average, refer to Chapter 6.

Example 1: (Area of Trapezoid)
❏ Calculate the area of a trapezoid with a large base of 5 ft, a small base of 2 ft, and a trapezoid height of 4 ft.

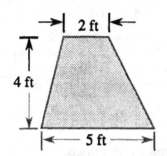

$$A = \frac{(b_1 + b_2)(h)}{2}$$

$$= \frac{(2 \text{ ft} + 5 \text{ ft})(4 \text{ ft})}{2}$$

$$= (3.5 \text{ ft})(4 \text{ ft})$$

$$= \boxed{\begin{array}{c} 14 \text{ sq ft} \\ \text{Surface Area} \end{array}}$$

❏ The area of a trapezoid is 40 sq ft. If the small base is 6 ft and the large base is 10 ft, what is the height of the trapezoid?

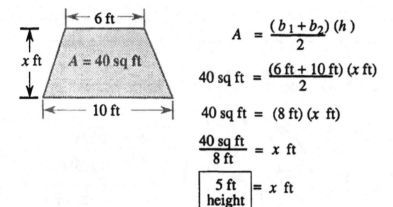

$$A = \frac{(b_1 + b_2)(h)}{2}$$

$$40 \text{ sq ft} = \frac{(6 \text{ ft} + 10 \text{ ft})(x \text{ ft})}{2}$$

$$40 \text{ sq ft} = (8 \text{ ft})(x \text{ ft})$$

$$\frac{40 \text{ sq ft}}{8 \text{ ft}} = x \text{ ft}$$

$$\boxed{\begin{array}{c} 5 \text{ ft} \\ \text{height} \end{array}} = x \text{ ft}$$

CALCULATING BASES AND HEIGHTS

The four variables in the trapezoid equation are area, (A), small base (b_1), large base (b_2), and height (h). If three of the variables are known, the fourth variable may be calculated.

AREA OF A CIRCLE

The easiest way to remember the equation for the area of a circle is to think of the circle as a "square with the corners cut off." The equation preferred for water and wastewater calculations is:

$$A_O = (0.785)(D^2)$$

where D is the diameter of the circle.* This equation is preferred over the more traditional equation

$$A = \pi r^2$$

because tanks and pipes have dimensions given in terms of diameters. Converting diameter to radius introduces an unnecessary possibility for error.

A CIRCLE IS A SQUARE WITH THE CORNERS CUT OFF

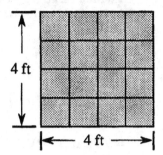

$$A = lw$$
$$= (4\ ft)(4\ ft)$$
$$= \boxed{\begin{array}{c}16\ sq\ ft \\ Surface\ Area\end{array}}$$

A circle is 78.5% the area of the same size square:

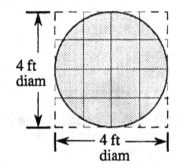

$$A = (0.785)(D^2)$$
$$= (0.785)(D)(D)$$
$$= (0.785)(4\ ft)(4\ ft)$$
$$= \boxed{\begin{array}{c}12\ sq\ ft \\ Surface\ Area\end{array}}$$

It is not necessary to construct a square around each circle to be calculated. Mathematically, the D^2 part of the equation represents the area of a square, while the 0.785 part of the equation is equivalent to "cutting the corners off the square."

Example 1: (Area of Circle)
❑ The diameter of a circle is 5 ft. What is its area?

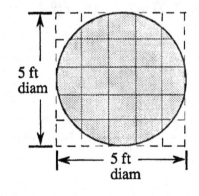

$$A = (0.785)(D^2)$$
$$= (0.785)(5\ ft)(5\ ft)$$
$$= \boxed{\begin{array}{c}20\ sq\ ft \\ Surface\ Area\end{array}}$$

---

* The diameter of a circle is any straight line drawn from one side of the circle THROUGH THE CENTER to the other side.

Example 2: (Area of Circle)
❏ What is the area of a circle when the diameter is 7 ft?

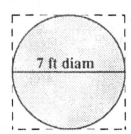

7 ft diam

$$A = (0.785)(D^2)$$
$$= (0.785)(7\text{ ft})(7\text{ ft})$$
$$= \boxed{\begin{array}{c}38\text{ sq ft}\\ \text{Surface Area}\end{array}}$$

Example 3: (Area of Circle)
❏ A circular clarifier has a diameter of 40 ft. What is the sq ft surface area of the clarifier?

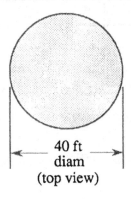

40 ft
diam
(top view)

$$A = (0.785)(D^2)$$
$$= (0.785)(40\text{ ft})(40\text{ ft})$$
$$= \boxed{\begin{array}{c}1256\text{ sq ft}\\ \text{Surface Area}\end{array}}$$

Example 4: (Area of Circle)
❏ If the area of a circle is 6358.5 sq ft, what is the diameter of the circle?

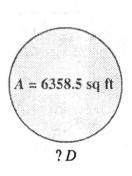

$A = 6358.5$ sq ft

? D

$$A = (0.785)(D^2)$$
$$6358.5\text{ sq ft} = (0.785)(D^2)$$
$$\frac{6358.5\text{ sq ft}}{0.785} = D^2$$
$$8100\text{ sq ft} = D^2$$
$$\sqrt{8100\text{ sq ft}} = D$$
$$\boxed{90\text{ ft}} = D$$

CALCULATING DIAMETERS

The two variables in calculating the area of a circle are area (A), and diameter (D). If one variable is known, the other may be calculated. This means that if you know the area of circle, you can calculate its diameter. Square roots are required to complete this type of problem; however, most calculators include the square root function ($\sqrt{}$) and can handle such a problem. (Refer to Powers and Square Roots, Chapter 13, for additional information concerning square roots.)

PRACTICE PROBLEMS: Areas 10.1: Basic Shapes

❑ Calculate the sq ft areas for the figures shown below:

1.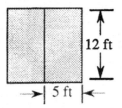

D = 80 ft

ANS_____

4.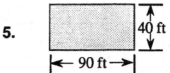

45 ft

(Square)

ANS_____

2.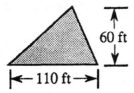

12 ft

5 ft

ANS_____

5.

40 ft

←— 90 ft —→

ANS_____

3.

60 ft

←— 110 ft —→

ANS_____

6.

←— 6 ft —→

4 ft

←— 4 ft —→

ANS_____

❑ Solve the following problems, as indicated.

7. The top of a tank has a surface area of 3150 sq ft. If the width of the tank is 35 ft, what is the length of the tank?

ANS_____

8. What is the diameter of a pipe whose cross-sectional area is 452.16 sq ft?

ANS_____

9. The cross-sectional area of a trapezoidal channel is 37.5 sq ft. If the top of the channel is 10 ft, and the depth of the channel is 5 ft, what is the measurement (in ft) across the bottom of the channel?

ANS_____

10. If the base of a triangle is 50 ft and the cross-sectional area is 375 sq ft, what is the height of the triangle?

ANS_____

10.2 AREAS—COMBINED SHAPES

> ### SUMMARY
>
> To calculate the area of a combined shape, calculate the area of each basic shape, then add the areas. Examples of three combined shapes typical to water and wastewater calculations are given below.
>
>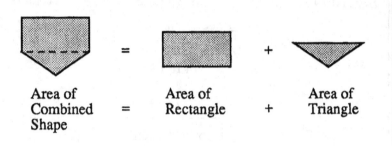
>
> | Area of Combined Shape | = | Area of Rectangle | + | Area of Triangle |
>
>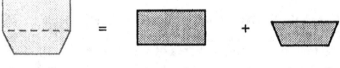
>
> | Area of Combined Shape | = | Area of Rectangle | + | Area of Trapezoid |
>
> Area of Combined Shape = Area of Rectangle + Area of 1/2 Circle

The four basic shapes in area calculations are the rectangle, triangle, trapezoid, and circle. More complex areas are often a combination of two or more of these basic shapes. Examples of some of the combined shapes you might encounter in water and wastewater calculations are shown to the left.

AREA OF COMBINED SHAPES

Sometimes area calculations involve a combination of two or more of the basic shapes described in Section 10.1. To calculate a combined area:

1. Calculate the area of each basic shape represented, and

2. Add the areas.

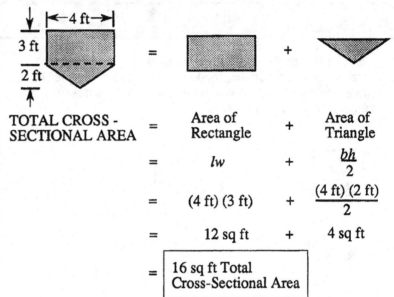

Example 1: (Areas—Combined Shapes)

❏ The cross-sectional area of a tank has dimensions as shown. Calculate the total cross-sectional area.

TOTAL CROSS - SECTIONAL AREA	=	Area of Rectangle	+	Area of Triangle
	=	lw	+	$\dfrac{bh}{2}$
	=	(4 ft) (3 ft)	+	$\dfrac{(4\ ft)\ (2\ ft)}{2}$
	=	12 sq ft	+	4 sq ft
	=	16 sq ft Total Cross-Sectional Area		

Example 2: (Areas—Combined Shapes)

❏ A channel has dimensions as shown below. Calculate the total cross-sectional area of the channel.

TOTAL CROSS - SECTIONAL AREA	=	Area of Rectangle	+	Area of 1/2 Circle
TOTAL CROSS - SECTIONAL AREA	=	lw	+	Area of 1/2 Circle
	=	(4 ft) (2 ft)	+	$\dfrac{(0.785)\ (4\ ft)\ (4\ ft)}{2}$
	=	8 sq ft	+	6.3 sq ft
	=	14.3 sq ft Total Cross-Sectional Area		

Example 3: (Areas—Combined Shapes)

❑ Calculate the cross-sectional area of the channel shown below.

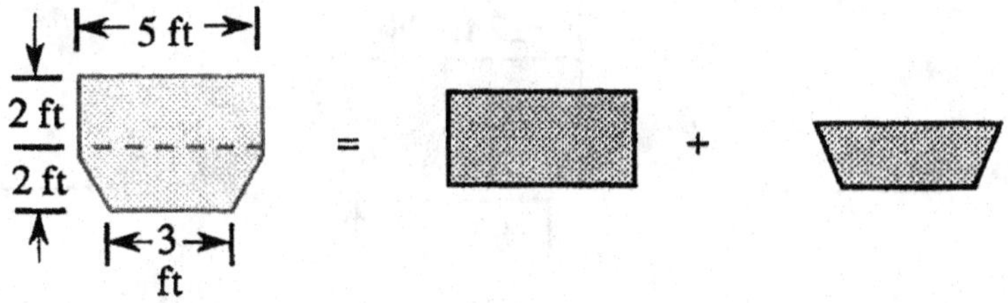

TOTAL CROSS - SECTIONAL AREA	=	Area of Rectangle	+	Area of Trapezoid
	=	lw	+	$\dfrac{(b_1 + b_2)(h)}{2}$
	=	$(5 \text{ ft})(2 \text{ ft})$	+	$\dfrac{(3 \text{ ft} + 5 \text{ ft})(2 \text{ ft})}{2}$
	=	10 sq ft	+	8 sq ft

$$= \boxed{\begin{array}{l}\text{18 sq ft Total} \\ \text{Cross-Sectional Area}\end{array}}$$

PRACTICE PROBLEMS 10.2: Areas—Combined Shapes

❏ Calculate the sq ft areas for the shaded portions of the figures shown below.

1.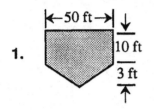

ANS_____

4.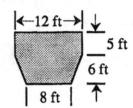

ANS_____

2.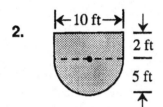

ANS_____

5.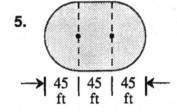

ANS_____

3.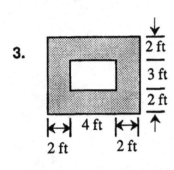

ANS_____

6.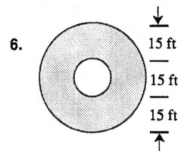

ANS_____

❏ Solve the following problems, as indicated.

7. The cross-section of a tank is shown below. If the total cross-sectional area is 776.6 sq ft, with dimensions as shown, what is the total depth of the tank measured from the center?

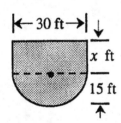

ANS_____

8. The cross-sectional area of the channel shown below is 42 sq ft. What is the width of the top of channel?

ANS_____

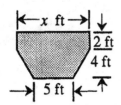

10.3 LATERAL AREAS

SUMMARY

1. To calculate the lateral area of a cylinder:

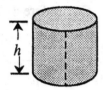

πD
(Circumference)

Based on the area of a rectangle:

$$A = lw$$

$$\boxed{L.\,A. = (\pi D)h}$$

2. To calculate the lateral area of a cone:

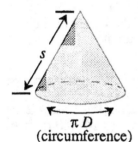

πD
(circumference)

Based on the area of a triangle:

$$A = \frac{bh}{2}$$

$$\boxed{L.\,A. = \frac{(\pi D)s}{2}}$$

3. To calculate the surface area of a sphere:

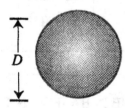

Related to the area of a circle:

$$A = (4)\,(\text{Area of Circle})$$

$$\boxed{L.\,A. = (4)(0.785)(D^2)}$$

The **total area** of an object is the area of the sides plus the area of the top and bottom (called "the bases"). The **lateral area** of an object is the area of the **sides only**.

Most calculations of area in water and wastewater math involve finding rectangular, triangular, trapezoidal, and circular areas.

Occasionally you may need to calculate the lateral area of three additional shapes: cylinders, cones, and spheres. This section describes the calculation of area for the sides of cylinders and cones and for spheres.

LATERAL AREA OF A CYLINDER

The lateral area of a cylinder is the **area of its sides only**. It does not include the area of its bases. To calculate the lateral area of a cylinder, imagine opening the cylinder so that it is flat, as shown to the right. The area is now a simple rectangle. Since the area of a rectangle is

$$A = lw,$$

the area of the "flattened cylinder" is also l (represented by πD) times w (represented by h).

LATERAL AREA OF A CONE

The equation for the lateral area of a cone (or pyramid) is derived from the equation for the area of a triangle:

$$A = \frac{bh}{2}$$

The b represents the total length of the base (in other words, its **circumference**) and the h is the height of the triangle on the face of the cone. (This is called its **slant height**.) The equation for the lateral area of a cone is therefore:

$$L.A. = \frac{(\pi D)(s)}{2}$$

To remember this equation, think of opening the cone. It forms an approximate triangle whose base is a circumference and whose height is the slant height, s, of the cone.

THE LATERAL AREA OF A CYLINDER IS BASED ON THE AREA OF A RECTANGLE

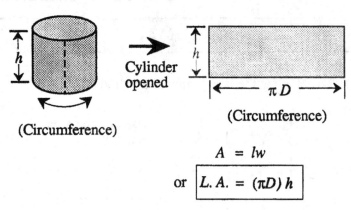

Cylinder opened

(Circumference)

(Circumference)

$$A = lw$$

or $\boxed{L.A. = (\pi D)\, h}$

Example 1: (Lateral Areas—Cylinder)
❑ The diameter of a tank is 40 ft. If the height of the tank is 10 ft, what is the lateral area of that tank?

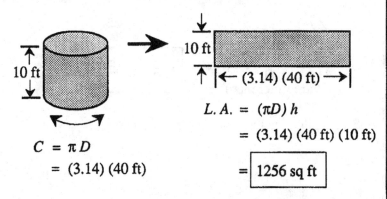

$$C = \pi D$$
$$= (3.14)(40 \text{ ft})$$

$$L.A. = (\pi D)\, h$$
$$= (3.14)(40 \text{ ft})(10 \text{ ft})$$
$$= \boxed{1256 \text{ sq ft}}$$

THE LATERAL AREA OF A CONE IS BASED ON THE AREA OF A TRIANGLE

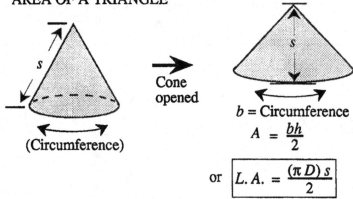

Cone opened

(Circumference)

$$b = \text{Circumference}$$
$$A = \frac{bh}{2}$$

or $\boxed{L.A. = \frac{(\pi D)\, s}{2}}$

Example 2: (Lateral Areas—Cone)
❑ The diameter of a cone is 6 ft. If the slant height of the cone is 5 ft. What is the lateral area of the cone?

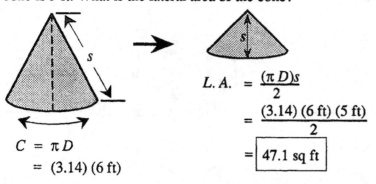

$$L.A. = \frac{(\pi D)s}{2}$$

$$= \frac{(3.14)\,(6\text{ ft})\,(5\text{ ft})}{2}$$

$$= \boxed{47.1 \text{ sq ft}}$$

$$C = \pi D$$
$$= (3.14)\,(6\text{ ft})$$

THE LATERAL AREA OF A SPHERE IS RELATED TO THE AREA OF A CIRCLE

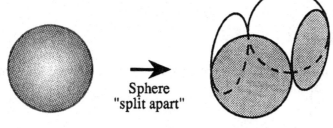

Sphere "split apart"

$$L.A. = (4)\,(\text{Area of Circle})$$

$$L.A. = \boxed{(4)\,(0.785)\,(D^2)}$$

LATERAL AREA OF A SPHERE

The lateral area of a sphere is a measure of the outside surface of that sphere.

A description of how the lateral area equation is derived is too lengthy to be presented here. However, a good way to remember the equation is that **the face of a sphere can be "split apart" into four equal circles:**

$$L.A. = (4)\,(\text{Area of Circle})$$

or $\boxed{L.A. = (4)\,(0.785)\,(D^2)}$

Example 3: (Lateral Areas—Sphere)
❑ The diameter of a sphere is 8 ft. What is the sq ft surface area of the sphere?

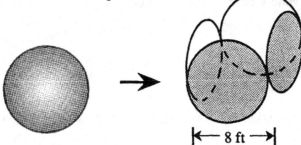

|← 8 ft →|

$$L.A. = (4)\,(0.785)\,(D^2)$$

$$= (4)\,(0.785)\,(8\text{ ft})\,(8\text{ ft})$$

$$= \boxed{201 \text{ sq ft}}$$

PRACTICE PROBLEMS 10.3: Lateral Areas

❑ Calculate the lateral area, in sq ft, for each of the figures given below.

1.

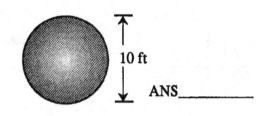

20 ft

$\leftarrow$ 80 ft $\rightarrow$
Diam

ANS_____

3.

10 ft

ANS_____

2.

20 ft

$\leftarrow$ 4 ft $\rightarrow$

ANS_____

4.

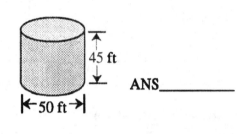

45 ft

$\leftarrow$ 50 ft $\rightarrow$

ANS_____

❑ Solve the following problems as indicated.

5. The cone portion of a tank must be painted. If the diameter of the cone is 50 ft and the slant height is 20 ft, how many sq ft must be painted?

ANS_____

6. A cylindrical tank with a diameter of 30 ft has a lateral area of 2355 sq ft. What is the height of the tank?

ANS_____

7. What is the approximate diameter of a sphere with a lateral area of 50 sq ft?

ANS_____

8. If the diameter of a cone is 8 ft and the lateral area is 75.36 sq ft, what is the slant height of the cone?

ANS_____

11 *Volume Measurement*

Complete and score the following skills test. Each section should be scored separately in the box provided to the right. For Section 11.1, a score of 8 or above indicates you are sufficiently strong in that concept. A score of 7 or below indicates a review of that section is advisable. For Section 11.2, a score of 4 or above indicates you are sufficiently strong in that concept. A score of 3 or below indicates a review of that section is advisable.

11.1 Volumes—Basic Shapes

Number Correct

❑ Calculate the cubic feet volume of the figures shown below.

1. 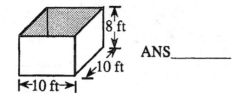   ANS_____

4. 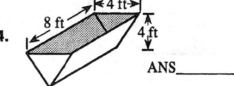   ANS_____

2. 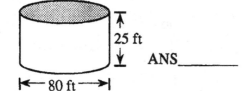   ANS_____

5. 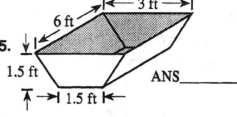   ANS_____

3. 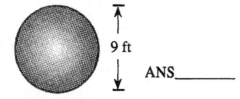   ANS_____

6. 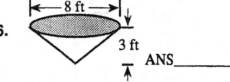   ANS_____

❑ Solve the following problems, as indicated.

7. A circular clarifier has a diameter of 40 ft. If the cu ft capacity of the clarifier is 18,840 cu ft, what is the maximum water depth in the tank?

ANS_____

8. A rectangular basin containing 26,400 cu ft of water has a length of 80 ft and is 30 ft wide. What is the depth of water in the tank?

ANS_____

(Continued)

9. A triangular trough 3 ft wide and 10 ft long contains 30 cu ft of water. What is the depth of the water in the trough.

ANS_____

10. What is the cu ft capacity of a 5-foot section of a 2-ft diameter pipe?

ANS_____

11.2 Volumes—Combined Shapes

❏ Calculate the cubic feet capacity for each figure shown below.

1.

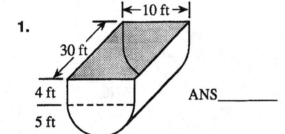

ANS_____

4.

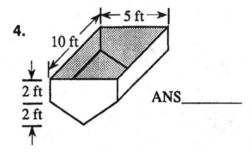

ANS_____

2.

ANS_____

5.

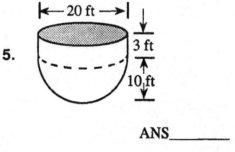

ANS_____

3.

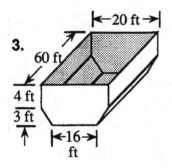

ANS_____

11.1 VOLUMES—BASIC SHAPES

SUMMARY

1. The general equation for most volume calculations is:

$$\text{Volume} = \begin{bmatrix} \text{Representative} \\ \text{Surface Area} \end{bmatrix} \begin{bmatrix} \text{Third} \\ \text{Dimension} \end{bmatrix}$$

2. The equations for the four basic volume shapes are as follows:

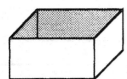

Rectangular Prism

$V = (lw)$ (3rd Dim.)

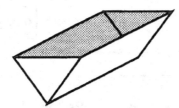

Triangular Prism

$V = \dfrac{(bh)}{2}$ (3rd Dim.)

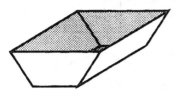

Trapezoidal Prism

$V = \dfrac{(b_1 + b_2)}{2} (h)$ (3rd Dim.)

Cylinder

$V = (0.785)(D^2)$ (3rd Dim.)

3. The equations for the volume of a cone and sphere are as follows:

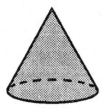

Cone

$V = \dfrac{1}{3}(0.785)(D^2)(h)$

Sphere

$V = \dfrac{2}{3}(0.785)(D^2)(D)$

A measurement of volume indicates the holding capacity of an object. A simple approach to most volume calculations is to remember that **volume is surface area times a third dimension**. The surface area used in these calculations must be the **representative surface**, the side that gives the object its basic shape. In the diagram to the left, the representative surface areas of a rectangle, triangle, trapezoid and circle have been illustrated. These are the four basic shapes most often found in water and wastewater calculations.

Calculations of the **cone and sphere volume do not use the general volume equation** since there is no "representative surface area." Both of these volume calculations, however, are based on the volume of a cylinder. The cone volume is one-third the volume of a cylinder and the sphere volume is two-thirds the volume of a special cylinder, as described in this section.

RECTANGULAR VOLUMES

The volume of a rectangular solid (sometimes called a rectangular prism) is calculated by multiplying the representative area (a rectangle) by the third dimension:

Volume = (Rep. Area) (3rd Dim)

Volume = $(lw)(h)$

Example 1: (Rectangular Volume)

❑ What is the cubic feet capacity of the tank shown below?

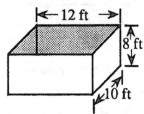

Volume = (Rep. Area) (3rd Dim.)

= $(lw)(h)$

= (12 ft) (10 ft) (8 ft)

= 960 cu ft

Example 2: (Rectangular Volume)

❑ Calculate the cubic feet volume of the tank shown below.

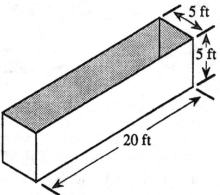

Volume = (Rep. Area) (3rd Dim.)

= $(lw)(h)$

= (20 ft) (5 ft) (5 ft)

= 500 cu ft

Example 3: (Rectangular Volume)
❑ A tank is 25 ft wide, 75 ft long, and contains water to a depth of 10 ft. How many cubic feet of water are in the tank?

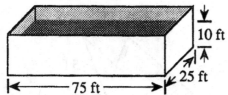

$$\text{Volume} = (\text{Rep. Area})\,(\text{3rd Dim.})$$
$$= (lw)\,(h)$$
$$= (75\text{ ft})\,(25\text{ ft})\,(10\text{ ft})$$
$$= \boxed{18{,}750\text{ cu ft}}$$

Example 4: (Rectangular Volume)
❑ A tank containing 1296 cu ft of water is 12 ft wide and 12 ft long. What is the depth of the water?

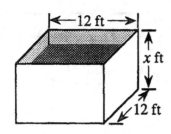

$V = 1296$ cu ft

$$\text{Volume} = (\text{Rep. Area})\,(\text{3rd Dim.})$$
$$= (lw)\,(h)$$
$$1296\text{ cu ft} = (12\text{ ft})\,(12\text{ ft})\,(x\text{ ft})$$
$$\frac{1296\text{ cu ft}}{(12\text{ ft})\,(12\text{ ft})} = x$$
$$\boxed{9\text{ ft}} = x$$

CALCULATING ONE OF THE DIMENSIONS

In some capacity or volume calculations, the volume will be known but one of the tank dimensions will be unknown. In these problems **begin as usual by writing the equation**, then fill in the information given and solve for the unknown value.*

---

* Refer to Chapter 2, Solving for the Unknown.

TRIANGULAR VOLUME

A triangular object, such as that shown in Example 4, is called a triangular prism (sometimes called a trough). Since the triangle gives the object its basic shape, **the triangle is the representative area**. Volume is therefore calculated as:

Volume = (Rep. Area) (3rd Dim.)

$$\text{Volume} = \frac{(bh)\,(\text{length})}{2}$$

Example 4: (Triangular Volume)
❑ Calculate the cubic feet capacity of the trough shown below.

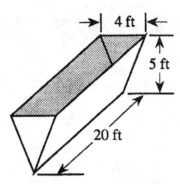

$$\begin{aligned}
\text{Volume} &= (\text{Rep. Area})\ (\text{3rd Dim.}) \\
&= \frac{(bh)\,(\text{length})}{2} \\
&= \frac{(4\ \text{ft})\ (5\ \text{ft})\ (20\ \text{ft})}{2} \\
&= \boxed{200\ \text{cu ft}}
\end{aligned}$$

Example 5: (Triangular Volume)
❑ Calculate the cubic feet capacity of the trough shown below.

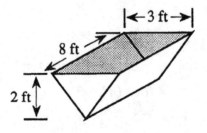

$$\begin{aligned}
\text{Volume} &= (\text{Rep. Area})\ (\text{3rd Dim.}) \\
&= \frac{(bh)\,(\text{length})}{2} \\
&= \frac{(3\ \text{ft})\ (2\ \text{ft})\ (8\ \text{ft})}{2} \\
&= \boxed{24\ \text{cu ft}}
\end{aligned}$$

Example 6: (Triangular Volume)
❑ The bottom portion of a tank is a triangular shape, as shown below. If the depth of the triangular portion is 5 ft, the width 10 ft and the length 40 ft, what is the cubic feet capacity of this portion of the tank?

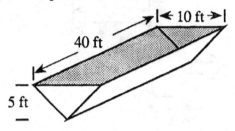

$$\text{Volume} = (\text{Rep. Area}) (\text{3rd Dim.})$$

$$= \frac{(bh)(\text{length})}{2}$$

$$= \frac{(10 \text{ ft}) (5 \text{ ft}) (40 \text{ ft})}{2}$$

$$= \boxed{1000 \text{ cu ft}}$$

Example 7: (Triangular Volume)
❑ The triangular tank shown below contains 168 cubic feet of water. If the depth of the water in the trough is 4 feet and the width of the trough is 3 feet, what is the length of the trough?

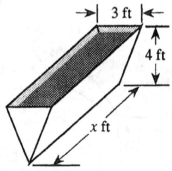

$$\text{Volume} = \frac{(bh)(\text{length})}{2}$$

$$168 \text{ cu ft} = \frac{(3 \text{ ft}) (4 \text{ ft}) (x \text{ ft})}{2}$$

$$\frac{(2) (168 \text{ cu ft})}{(3 \text{ ft}) (4 \text{ ft})} = x$$

$$\boxed{28 \text{ ft}} = x$$

CALCULATING ONE OF THE DIMENSIONS

In volume calculations such as these, there are four variables: base (b), height (h), length (l), and cubic feet volume (V).

$$\boxed{V = \frac{(bh)(l)}{2}}$$

When any three of these variables are known, the fourth variable may be calculated as shown in Example 7.

TRAPEZOIDAL VOLUME

A trapezoidal volume such as that shown in Example 8 is calculated by multiplying the representative area (a trapezoid) by the third dimension:

Volume = (Rep. Area) (3rd Dim.)

$$\text{Volume} = \frac{(b_1 + b_2)\,(h)\,(\text{length})}{2}$$

Example 8: (Trapezoidal Volume)

❑ A trapezoidal tank has dimensions as shown below. If the water depth in the tank is 4 ft, what is the cubic feet volume of water in the tank?

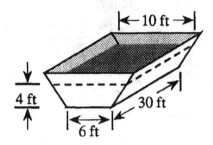

$$\begin{aligned}
\text{Volume} &= (\text{Rep. Area}) \, (\text{3rd Dim.})\\[4pt]
&= \frac{(b_1 + b_2)\,(h)\,(\text{3rd Dim.})}{2}\\[4pt]
&= \frac{(6\text{ ft} + 10\text{ ft})\,(4\text{ ft})\,(30\text{ ft})}{2}\\[4pt]
&= \boxed{960 \text{ cu ft}}
\end{aligned}$$

Example 9: (Trapezoidal Volume)

❑ Calculate the cu ft capacity of the tank shown below.

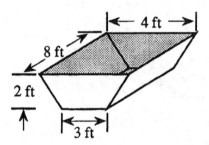

$$\begin{aligned}
\text{Volume} &= (\text{Rep. Area}) \, (\text{3rd Dim.})\\[4pt]
&= \frac{(b_1 + b_2)\,(h)\,(\text{3rd Dim.})}{2}\\[4pt]
&= \frac{(3\text{ ft} + 4\text{ ft})\,(2\text{ ft})\,(8\text{ ft})}{2}\\[4pt]
&= \boxed{56 \text{ cu ft}}
\end{aligned}$$

Example 10: (Trapezoidal Volume)
❑ The trapezoidal tank shown below has a capacity of 420 cubic feet. If the long base of the trapezoid is 8 ft, the short base is 6 ft, and the length of the tank is 20 ft, what is the maximum water depth possible in the tank?

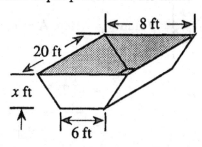

$$\text{Volume} = (\text{Rep. Area})\,(\text{3rd Dim.})$$

$$= \frac{(b_1 + b_2)\,(h)\,(\text{3rd Dim.})}{2}$$

$$420\text{ cu ft} = \frac{(6\text{ ft} + 8\text{ ft})\,(x\text{ ft})\,(20\text{ ft})}{2}$$

$$420\text{ cu ft} = (7\text{ ft})\,(x\text{ ft})\,(20\text{ ft})$$

$$\boxed{3\text{ ft}} = x$$

CALCULATING ONE OF THE DIMENSIONS

Trapezoidal volume calculations have five variables: the shorter base (b_1), the longer base (b_2), the height (h), the third dimension, and the volume (V). Given information about any four of these variables, the fifth variable may be calculated, as shown in Examples 10 and 11.

Example 11: (Trapezoidal Volume)
❑ A certain trapezoidal tank contains 120 cu ft of water. What is the length of the tank given the water depth and dimensions as shown below?

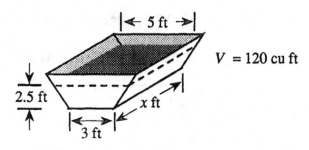

$V = 120$ cu ft

$$\text{Volume} = (\text{Rep. Area})\,(\text{3rd Dim.})$$

$$120\text{ cu ft} = \frac{(3\text{ ft} + 5\text{ ft})\,(2.5\text{ ft})\,(x\text{ ft})}{2}$$

$$120\text{ cu ft} = (4\text{ ft})\,(2.5\text{ ft})\,(x\text{ ft})$$

$$\boxed{12\text{ ft}} = x$$

CYLINDRICAL VOLUME

The top and bottom of a cylinder (the bases) are circles. Therefore in calculating volume the "representative surface area" is a circle:

Volume = (Rep. Area) (3rd Dim.)

Volume = $(0.785)(D^2)(h)$

Example 12: (Cylindrical Volume)
❑ Calculate the cubic feet capacity of the tank shown below.

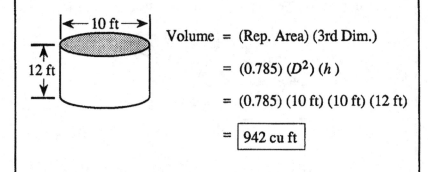

Volume = (Rep. Area) (3rd Dim.)

= $(0.785)(D^2)(h)$

= (0.785) (10 ft) (10 ft) (12 ft)

= 942 cu ft

Example 13: (Cylindrical Volume)
❑ The diameter of a tank is 60 ft and the height is 25 ft. What is the cubic feet capacity of that tank?

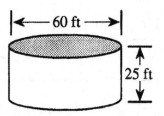

Volume = (Rep. Area) (3rd Dim.)

= $(0.785)(D^2)(h)$

= (0.785) (60 ft) (60 ft) (25 ft)

= 70,650 cu ft

Example 14: (Cylindrical Volume)

❑ The diameter of a tank is 80 ft. If the water depth in the tank is 12 ft, how many cubic feet of water are in the tank?

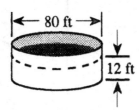

$$\text{Volume} = (\text{Rep. Area}) (\text{3rd Dim.})$$
$$= (0.785) (D^2) (h)$$
$$= (0.785) (80 \text{ ft}) (80 \text{ ft}) (12 \text{ ft})$$
$$= \boxed{60{,}288 \text{ cu ft}}$$

Example 15: (Cylindrical Volume)

❑ A tank holding 127,170 cubic feet of water has water to a depth of 20 ft. What is the diameter of the tank?

Volume = 127,170 cu ft

$$\text{Volume} = (\text{Rep. Area}) (\text{3rd Dim.})$$
$$127{,}170 \text{ cu ft} = (0.785) (D^2) (20 \text{ ft})$$
$$\frac{127{,}170 \text{ cu ft}}{(0.785) (20 \text{ ft})} = D^2$$
$$8100 = D^2$$
$$\sqrt{8100} = D$$
$$\boxed{90 \text{ ft}} = D$$

CALCULATING THE DIAMETER OR HEIGHT

There are three variables in cylinder volume calculations: Volume (V), Diameter (D), and height (h). If you have dimensions for **any two of these variables**, you can calculate the value of the third variable using the same volume equation.

First, fill in the given information, then solve for the unknown value.* If the unknown variable is the diameter, as shown in Example 15, you will have to calculate a square root to find the answer. However, you are not expected to calculate the square root by hand. Simply use the square root key on your pocket calculator to find the square root indicated.**

---

* Refer to Chapter 2, Solving for the Unknown Value.
** For a review of square roots, refer to Chapter 13, "Powers, Roots, and Scientific Notation."

CONE VOLUME

Compare the volumes of the cylinder and cone shown to the right. Given equal diameters and heights, the volume of the cone is exactly one-third the volume of the cylinder:

$$V = \frac{1}{3} \text{ (Vol. of Cylinder)}$$

$$\boxed{V = \frac{1}{3} (0.785) (D^2) (h)}$$

THE VOLUME OF A CONE IS 1/3 THE VOLUME OF THE CYLINDER

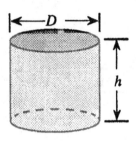

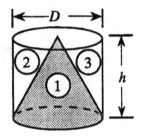

Volume of a Cylinder

$$V = \text{(Rep. Area) (3rd Dim.)}$$

$$= (0.785) (D^2) (h)$$

Volume of a Cone

$$V = \frac{1}{3} \text{(Vol. of Cylinder)}$$

$$= \frac{1}{3} (0.785) (D^2) (h)$$

Example 16: (Cone Volume)
❑ Calculate the cubic feet volume of the cone shown below.

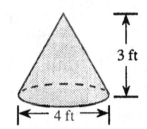

$$\text{Volume} = \frac{1}{3} \text{ (Volume of Cylinder)}$$

$$= \frac{1}{3} (0.785) (D^2) (h)$$

$$= \frac{(0.785) (4 \text{ ft}) (4 \text{ ft}) (3 \text{ ft})}{3}$$

$$= \boxed{12.6 \text{ cu ft}}$$

Example 17: (Cone Volume)

❏ The bottom portion of a tank is a cone. If the diameter is 50 ft and the height (or depth in this case) is 3 ft, what is the cubic feet capacity of the cone portion of the tank?

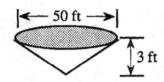

|← 50 ft →|

3 ft

$$\text{Volume} = \frac{1}{3}\ (\text{Volume of Cylinder})$$

$$= \frac{1}{3}\ (0.785)\ (D^2)\ (h)$$

$$= \frac{(0.785)\ (50\ \text{ft})\ (50\ \text{ft})\ (3\ \text{ft})}{3}$$

$$= \boxed{1962.5\ \text{cu ft}}$$

CALCULATING DIAMETER OR HEIGHT

Cone volume calculations have three variables: the diameter (D), the height (h), and the volume (V). If you know the values of **any two of the variables**, you will be able to calculate the value of the third variable.

Example 18: (Cone Volume)

❏ The volume of the cone shown below is 1884 cubic feet. If the cone has a diameter of 60 ft, what is the height of the cone?

|← 60 ft →|

$V = 1884$ cu ft

x ft

$$\text{Volume} = \frac{1}{3}\ (\text{Volume of Cylinder})$$

$$1884\ \text{cu ft} = \frac{1}{3}\ (0.785)\ (D^2)\ (h)$$

$$1884\ \text{cu ft} = \frac{(0.785)\ (60\ \text{ft})\ (60\ \text{ft})\ (x\ \text{ft})}{3}$$

$$\frac{(3)\ (1884\ \text{cu ft})}{(0.785)\ (60\ \text{ft})\ (60\ \text{ft})} = x$$

$$\boxed{2\ \text{ft}} = x$$

SPHERE VOLUME

In the previous section we learned that the volume of a cone within a cylinder is always 1/3 the volume of that cylinder. This is true regardless of the size of the cylinder—tall and narrow or short and squat.

The volume of a sphere is 2/3 the volume of the cylinder. However, **this is true for only one type of cylinder**—a cylinder whose height and diameter are equal.

THE VOLUME OF A SPHERE IS 2/3 THE VOLUME OF A SPECIAL CYLINDER

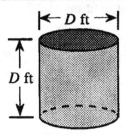

The special cylinder has a height equal to its diameter.

$$V = \text{(Rep. Area) (3rd Dim.)}$$

$$= (0.785)\,(D^2)\,(D)$$

$$= (0.785)\,(D^3)$$

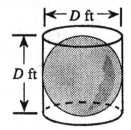

The volume of the sphere is 2/3 the volume of the special cylinder:

$$V = \frac{2}{3}\,\text{(Rep. Area) (3rd Dim.)}$$

$$= \frac{2}{3}\,(0.785)\,(D^2)\,(D)$$

$$= \frac{2}{3}\,(0.785)\,(D^3)$$

Example 19: (Sphere Volume)
❑ Calculate the cubic feet volume of the sphere shown below.

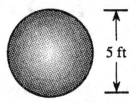

5 ft

$$V = \frac{2}{3}\,\text{(Volume of Special Cylinder)}$$

$$= \frac{2}{3}\,(0.785)\,(D^2)\,(D)$$

$$= \frac{2}{3}\,(0.785)\,(5\text{ ft})\,(5\text{ ft})\,(5\text{ ft})$$

$$= \boxed{65.4 \text{ cu ft}}$$

Example 20: (Sphere Volume)
❑ The diameter of a sphere is 20 ft. What is the cubic feet volume of the sphere?

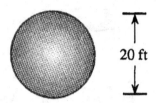

$$V = \frac{2}{3} \text{ (Volume of Special Cylinder)}$$

$$= \frac{2}{3} (0.785) (D^2) (D)$$

$$= \frac{2}{3} (0.785) (20 \text{ ft}) (20 \text{ ft}) (20 \text{ ft})$$

$$= \boxed{4187 \text{ cu ft}}$$

Example 21: (Sphere Volume)
❑ The bottom portion of a cylindrical tank is a half-sphere. If the diameter of the tank is 6 ft, what is the cubic feet volume of the half sphere?

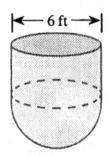

6 ft

$$V = \frac{1}{2} \left[\frac{2}{3} \text{ (Volume of Special Cylinder)} \right]$$

$$= \frac{1}{2} \left[\frac{2}{3} (0.785) (D^2) (D) \right]$$

$$= \frac{1}{3} (0.785) (6 \text{ ft}) (6 \text{ ft}) (6 \text{ ft})$$

$$= \boxed{56.5 \text{ cu ft}}$$

In this problem you are calculating the volume of only **half a sphere**. Therefore, the usual sphere equation is multiplied by 1/2. In step 3 of the calculation, notice that the fractions 1/2 and 2/3 have been multiplied together (resulting in 2/6), then reduced to 1/3.

PRACTICE PROBLEMS 11.1: Volumes—Basic Shapes

❏ Calculate the cubic feet volume of the figures shown below:

1.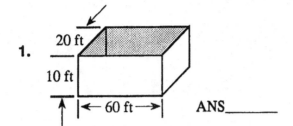

20 ft

10 ft

← 60 ft →

ANS_____

2.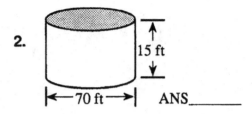

15 ft

← 70 ft →

ANS_____

3.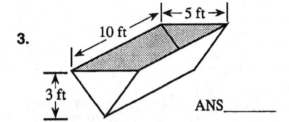

← 5 ft →

10 ft

3 ft

ANS_____

4.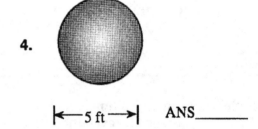

← 5 ft →

ANS_____

5.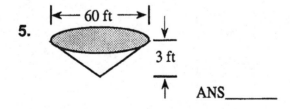

← 60 ft →

3 ft

ANS_____

6.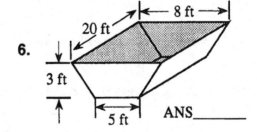

← 8 ft →

20 ft

3 ft

5 ft

ANS_____

PRACTICE PROBLEMS 11.1: Volumes—Basic Shapes (Continued)

❑ Complete the following problems, as indicated.

7. A clarifier has a diameter of 50 ft. If the depth of water in the clarifier is 15 ft, how many cubic feet of water are in the clarifier?

ANS_____

8. A rectangular basin 25 ft wide and 75 ft long contains 28,125 cubic feet of water. What is the depth of the water in the tank?

ANS_____

9. The bottom portion of a tank is a triangular prism. If the base of the triangle is 20 ft, the depth of the triangular part of the tank is 3 ft and the length of the tank is 60 ft, how many cubic feet of water will this triangular portion hold?

ANS_____

10. What is the cubic feet capacity of a 2000-ft section of 18-inch-diameter pipe?

ANS_____

NOTES:

11.2 VOLUMES—COMBINED SHAPES

SUMMARY

Some of the common combined shapes found in volume calculations are shown below:

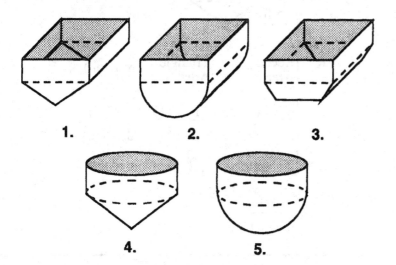

1. **2.** **3.**

4. **5.**

To calculate the total volume of combined shapes, there are two possible approaches. The first approach listed below is effective only when you can determine a "representative surface area," such as for diagrams 1-3. The second approach may be used whether or not a "representative surface area" can be determined.

1. Calculate the total cross-sectional area then multiply by the third dimension. For example, using this method for the figure shown in diagram 1:

    Total Volume = (Total Area) (3rd Dimension)

    $$= \left[\begin{array}{c} \text{Rectangle} + \text{Triangle} \\ \text{Area} \qquad \text{Area} \end{array}\right] \text{(3rd Dim.)}$$

    $$= \left[\; \blacksquare \;+\; \blacktriangledown \;\right] \text{(3rd Dim.)}$$

2. Calculate the volume of each basic shape, then total these volumes:

    $$\text{Total Volume} = \begin{array}{c} \text{Volume of} \\ \text{Rectangular} \\ \text{Shape} \end{array} + \begin{array}{c} \text{Volume of} \\ \text{Triangular} \\ \text{Shape} \end{array}$$

Volume calculations sometimes involve move complex shapes, such as those shown to the left. The method of calculating the volume depends on the type of combined shape, as described in the summary.

COMPLEX SHAPES WITH REPRESENTATIVE SURFACE AREAS

When calculating the volume of complex shapes with "representative surface areas," such as those shown in examples 1 and 2, two different methods may be used:

1. Calculate the total cross-sectional area, then multiply by the third dimension (see Example 1); or

2. Calculate the volume of each basic shape, then total the volumes (see Example 2).

Either method may be used; however, Method 1 may tend to result in fewer errors. Using Method 2, it is quite common to forget the last step of adding all volumes to obtain the total volume.

Example 1: (Volume—Combined Shapes)
❑ Calculate the cubic feet volume of the tank shown below.

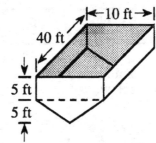

$$V = \left[\begin{array}{c} \text{Rectangle} \\ \text{Area} \end{array} + \begin{array}{c} \text{Triangle} \\ \text{Area} \end{array} \right] (\text{3rd Dim.})$$

$$= \left[(5\text{ ft})(10\text{ ft}) + \frac{(10\text{ ft})(5\text{ ft})}{2} \right] (40\text{ ft})$$

$$= \left[50\text{ sq ft} + 25\text{ sq ft} \right] (40\text{ ft})$$

$$= (75\text{ sq ft})(40\text{ ft})$$

$$= \boxed{3000\text{ cu ft}}$$

Example 2: (Volume—Combined Shapes)
❑ Calculate the volume of the tank shown below.

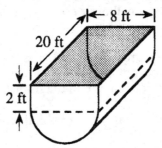

$$V = \begin{array}{c} \text{Volume of} \\ \text{Rectangle} \end{array} + \begin{array}{c} \text{Volume of} \\ \text{Half Cylinder} \end{array}$$

$$= (8\text{ ft})(2\text{ ft})(20\text{ ft}) + \frac{(0.785)(8\text{ ft})(8\text{ ft})(20\text{ ft})}{2}$$

$$= 320\text{ cu ft} + 502\text{ cu ft}$$

$$= \boxed{822\text{ cu ft}}$$

Example 3: (Volume—Combined Shapes)
❑ Given the diagram below, what is the cubic feet volume of the tank?

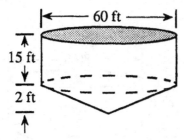

$$V = \text{Volume of Cylinder} + \text{Volume of Cone}$$

$$= (0.785)\,(60\text{ ft})\,(60\text{ ft})\,(15\text{ ft}) + \frac{(0.785)\,(60\text{ ft})\,(60\text{ ft})\,(2\text{ ft})}{3}$$

$$= 42{,}390 \text{ cu ft} + 1884 \text{ cu ft}$$

$$= \boxed{44{,}274 \text{ cu ft}}$$

COMPLEX SHAPES WITHOUT REPRESENTATIVE SURFACE AREAS

Certain complex shapes do not have representative surface areas. When this is the case, the total volume is determined by calculating the volumes of the basic shapes, then adding these volumes together. Examples 3 and 4 illustrate this type of calculation.

Example 4: (Volume—Combined Shapes)
❑ With dimensions as shown, what is the cubic feet capacity of the tank? (Assume the bottom portion of the tank is a half-sphere.)

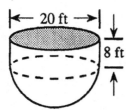

$$V = \text{Volume of Cylinder} + \text{1/2 Volume of a Sphere}$$

$$= (0.785)\,(20\text{ ft})\,(20\text{ ft})\,(8\text{ ft}) + \frac{\dfrac{(2)\,(0.785)\,(20\text{ ft})\,(20\text{ ft})\,(20\text{ ft})}{3}}{2}$$

$$= 2512 \text{ cu ft} + 2093 \text{ cu ft}$$

$$= \boxed{4605 \text{ cu ft}}$$

PRACTICE PROBLEMS 11.2: Volumes—Combined Shapes

❑ Calculate the cubic feet volume of the figures shown below.

1.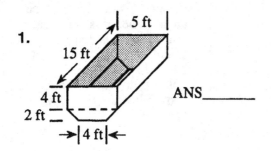

ANS_____

2.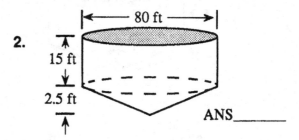

ANS_____

3.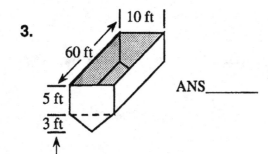

ANS_____

4.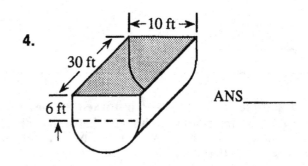

ANS_____

5.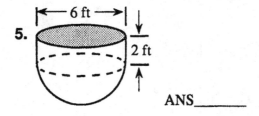

ANS_____

12   *Scales and Graphs*

Complete and score the following skills test. Each section should be scored separately in the box provided to the right. For Section 12.1, a score of 4 or above indicates that you are sufficiently strong in that concept. A score of 3 or below indicates a review of that section is advisable. For Section 12.2, a score of 8 or above indicates that you are sufficiently strong in that concept. A score of 7 or below indicates a review of that section is advisable.

12.1 Reading Scales

Number
Correct

❑ In the following problems, write the fraction (in reduced form) for each point on the scales given below.

1.

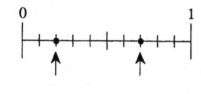

2.

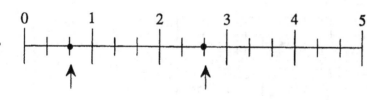

3.

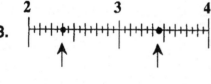

4.

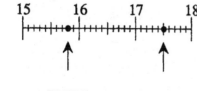

5.

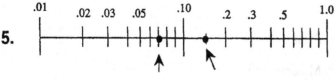

12.2 Types of Graphs

❑ Answer the questions given for each graph below.

1. During which months was the average monthly precipitation greater than 4 inches?

ANS_____

2. What was the approximate rainfall, in inches, during the wettest month?

ANS_____

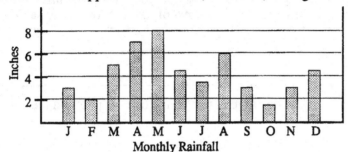

Monthly Rainfall

3. According to the graph shown below, chart approximately how many pounds of chlorine were used on Saturday?

ANS_____

4. On which two days did peak chlorine use occur?

ANS_____

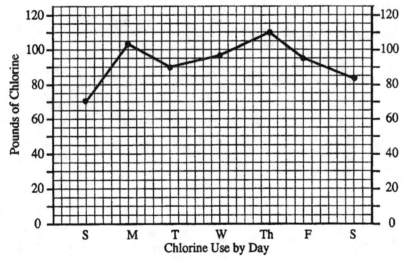

Chlorine Use by Day

5. What percent of the district budget is spent on construction and repair costs?

ANS_____

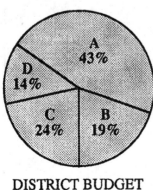

DISTRICT BUDGET

Key:

A = Construction Costs

B = Salaries

C = Repair Costs

D = Miscellaneous Expenses

12.2 Types of Graphs—Cont'd

❏ Use the nomograph below to answer tquestions 6 and 7.

6. What is the headloss for a 2-inch diameter pipeline flowing at 10 gpm?

ANS_____

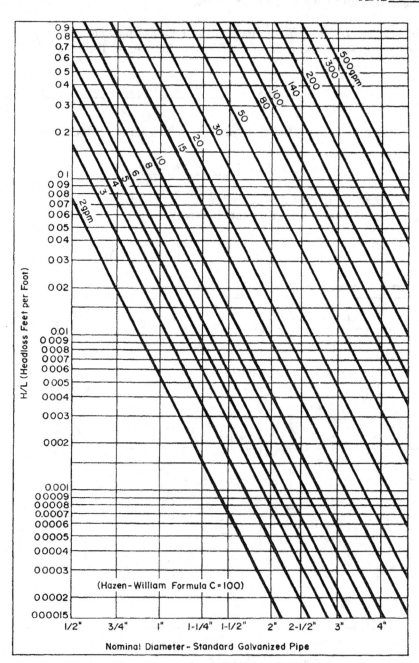

Head Loss Versus Pipe Size
(Source: EPA Manual of Individual Water Supply Systems, Rev. 1974)

12.2 Types of Graphs—Cont'd

7. What is the headloss for a 3-inch diameter pipeline flowing at a rate of 140 gpm? (Use the nomograph on the previous page.)

ANS_____

❏ Answer questions 8-10 using the graph shown below.

8. The alkalinity of a water is 150 mg/*L* and its pH is 8.0. Does this water fall in the zone of carbonate deposits, the zone of $CaCO_3$ equilibrium, or the zone of serious corrosion?

ANS_____

9. The alkalinity of a water is 225 mg/*L* and its pH is 7.3. In what zone does this water fall?

ANS_____

10. The alkalinity of a water is 175 mg/*L*. What range of pH is necessary so the water results in neither corrosion nor carbonate deposits?

ANS_____

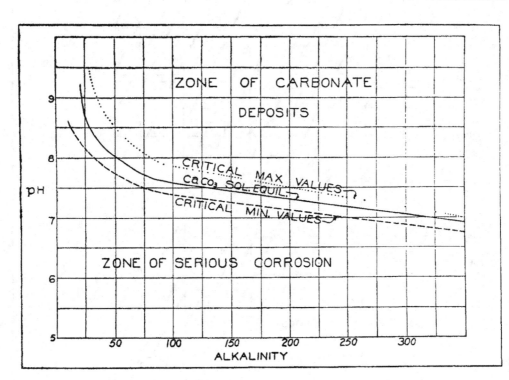

Relationship of pH vs Alkalinity
(Source: New York Manual of Instruction for Water Treatment Plant Operators)

12.1 READING SCALES

SUMMARY

1. To determine points on a scale, you must know **both the whole number and fraction** (if any) represented by that point.

2. The **size of the tick marks** helps indicate the fractional portion of a number.

3. When a point falls between two tick marks, **add new tick marks** to determine the correct fraction.

4. Arithmetic scales have equal divisions; whereas, logarithmic scales are divided unevenly according to logarithmic values.

POINTS ON A SCALE

Reading scales is an essential skill in understanding and interpreting data. The skills used in determining points on a scale are the same as those for comparing and ranking fractions and mixed numbers.*

To name the points on a scale, you will have to know the whole number and the fraction represented by the point:

1. Determine the whole number indicated.

2. Determine the fraction. The **denominator** is determined by counting the number of line segments into which each whole number has been divided. The **numerator** is determined by counting the number of line segments from the beginning of the whole number to the arrow.

Example 1: (Reading Scales)
❑ Give the fraction or mixed number for each point indicated by the arrows. Show all fractions in lowest terms.

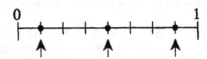

Step 1—Determine the whole number.
Since the arrows fall on the line before the number one, the whole number represented is 0. When zero is part of a mixed number, it is omitted and only the fraction is written. For example:

$$0\frac{1}{2} = \frac{1}{2}$$

Step 2—Determine the fraction.
The line between 0 and 1 has been divided into eight segments; therefore, the denominator of the fraction is 8. To find the numerator of each point, count the number of line segments (counting from left to right) until you reach the arrows.

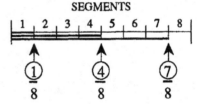

The three points are represented by the fractions 1/8, 4/8 and 7/8. Only 4/8 needs to be reduced:

$$\frac{4 \div 2}{8 \div 2} = \boxed{\frac{1}{2}}$$

The answer to this problem is 1/8, 1/2, and 7/8.

Example 2: (Reading Scales)
❑ Give the fraction or mixed number for each point indicated by the arrows. Show all fractions in lowest terms.

0 — 1

$$\boxed{\frac{5}{32}} \qquad \frac{18}{32} = \boxed{\frac{9}{16}} \qquad \frac{30}{32} = \boxed{\frac{15}{16}}$$

* For a discussion of fractions and mixed numbers, refer to Chapter 3.

Example 3: (Reading Scales)
❏ Give the fraction or mixed number for each point indicated on the scale. Reduce fractions to lowest terms.

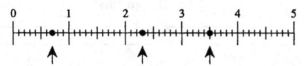

The scales given in Examples 1 and 2 were scales between 0 and 1. Most scales include several whole numbers, as shown in Example 3.

The first arrow indicates a number less than one. Therefore the point will be represented by a fraction only.

The arrow is pointing to the seventh mark, beginning at zero and counting to the right.

The denominator of the fraction is 10 since the distance between any two whole numbers has been broken into 10 segments.

The second arrow falls between the whole numbers 2 and 3. The number will be greater than two (because it has passed the 2 mark) but less than three (because the arrow has not yet reached the 3 mark). In fact, the number is 2 and a fraction:

$$2\frac{3}{10}$$

The third arrow falls between 3 and 4. It will be a mixed number of 3 and a fraction:

$$3\frac{5}{10} = \boxed{3\frac{1}{2}}$$

Example 4: (Reading Scales)
❏ Give the fraction or mixed number for each point indicated on the scale. Reduce fractions to lowest terms.

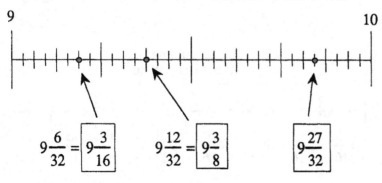

$$9\frac{6}{32} = \boxed{9\frac{3}{16}} \qquad 9\frac{12}{32} = \boxed{9\frac{3}{8}} \qquad \boxed{9\frac{27}{32}}$$

SHORTCUT IN READING SCALES

In Examples 1-3, the denominators of the fractions were determined by counting the total number of divisions or segments between two whole numbers. However, it is not always necessary to count all the segments. Scales are often marked in a way that enables a shortcut to be used.

In the shortcut method, it is important to **pay attention to the size of the tick marks.** The longest marks are used to mark off the whole numbers (9 and 10 in the scale to the right).

The next longest mark is the **halfway (1/2) mark.** An arrow pointing to the mark would be read as 9-1/2. In such a case, it would not be necessary to count all segments shown in the bottom scale.

The next longest marks indicate the **quarter (1/4) marks.** If an arrow were pointing to any of these marks, the answer could be obtained directly, without counting all 16 segments.

The next level of marks indicate the **eighth (1/8) marks.** Note that the second arrow in Example 4 was pointing to one of the eighth divisions. Therefore the answer 9-3/8 could have been obtained directly rather than finding 9-12/32 and reducing to 9-3/8.

The **sixteenth (1/16) marks** are slightly smaller than the eighth marks. The first arrow in Example 4 was pointing to 9-3/16.

Scale given in Example 4:

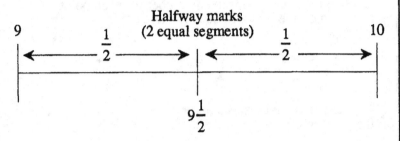

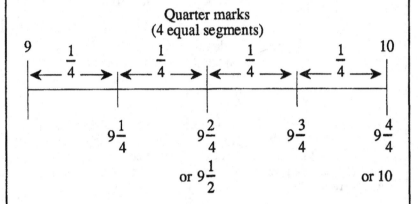

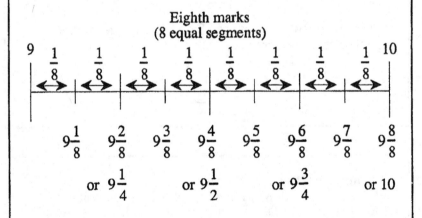

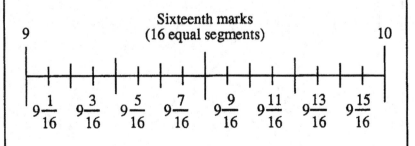

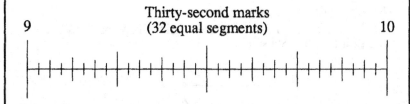

Thirty-second marks
(32 equal segments)

9   10

Note: a fast way to find out how many segments a line has been divided into. Count the segments to the halfway mark, then multiply by 2.

The smallest mark on this scale is represented by **thirty-second (1/32) marks**. Counting all marks, there are 32 equal line segments between 9 and 10. The third arrow in Example 4 was pointing to the smallest mark. When an arrow points to the smallest mark, it will not be able to be reduced any further. (The answer for the third arrow was 9-27/32.)

Example 5: (Reading Scales)
❑ Give the fractions or mixed numbers that correspond to each point shown on the scale. Reduce fractions to lowest terms.

21   22   23

Counting the segments between whole numbers, indicates that the smallest-sized segments are sixteenths. However in this problem we will use the shortcut method.

Arrow 1—The first arrow points to an eighth mark (21-3/8):

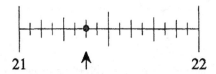

21   22

Arrow 2—The second arrow points to the smallest mark—a sixteenth (21-11/16):

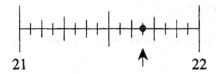

21   22

Arrow 3—The third arrow points to a quarter mark (22-3/4).

22   23

POINTS BETWEEN SCALE MARKINGS

What happens when the point you must read falls midway between two scale markings? How do you determine the correct fraction or mixed number for that point?

When a point falls midway between two tick marks, you simply change the existing scale:

1. **Add new tick marks** to the current scale. The new marks should be smaller than any other marks and placed between existing marks. The point in question should now fall directly on a new tick mark. (When the point does not fall halfway between points, it is sometimes necessary to add more than one set of tick marks, as shown in Example 7.)

2. **Name the point** as described in Examples 1-5.

Example 6: (Reading Scales)
❑ Determine the mixed number which corresponds with the point on the scale below.

To determine the name of the point indicated by the arrow, first add new tick marks between each existing mark: (Make the new marks smaller than other marks.)

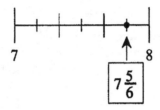

$$7\frac{5}{6}$$

The point is named as described in previous examples.

Example 7: (Reading Scales)
❑ Approximately what mixed number corresponds to the position marked by the arrow on the scale below?

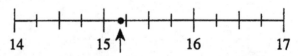

The arrow falls about three-fourths of the way from the 15 mark to the next mark. Therefore, two additional levels of tick marks will have to be added—the first halfway between each quarter mark; the second halfway between those marks. The scale would then be:

Now the arrow points directly to a tick mark. There are now 16 segments between 15 and 16. The number corresponding to the position of the arrow is:

$$15\frac{3}{16}$$

Example 8: (Reading Scales)
❏ What number is represented by the point shown on the scale below?

The current scale is in tenths (ten equal segments between whole numbers). To name the point given on the scale, new tick marks should be added.

The scale now has twice as many segments between whole numbers. Each segment is now 1/20 instead of 1/10. The point now lies on a tick mark:

$$50\frac{5}{20} = \boxed{50\frac{1}{4}}$$

DECIMAL SCALES

"Deci" is a prefix meaning tenth. Thus, decimal scales divide segments into **ten parts**. In Example 7, the scales have been divided according to the English system of measurements— halves, fourths, eighths, etc. A decimal scale is given in Example 8. And although the scale in Example 9 has been simplified to fifths, it is essentially a decimal scale as well.

Example 9: (Reading Scales)
❏ What number corresponds to the position marked by the arrow shown on the scale below?

There are five segments from 21 to 22. Since the arrow falls on the third tick mark, the number that corresponds to the position of the arrow is:

$$\boxed{21\frac{3}{5}}$$

READING LOGARITHMIC SCALES

The scales described thus far are **arithmetic scales**—scales with equal divisions. Another type of scale, **a logarithmic scale**, is used frequently for engineering data. These scales do not have equal divisions between points since the space is divided according to logarithmic values.*

Logarithmic scales can have as many as three levels of graduations, called primary, secondary, and tertiary graduations.

The **primary graduations** or tick marks divide the space into nine segments, beginning from one power of ten** to another. For example, the scale might range from 1 to 10, from 10 to 100, from 0.01 to 0.1, etc., as illustrated in the example at the top of the page.

Regardless of the beginning and ending numbers, **primary graduations divide the scale into nine segments**, with spaces decreasing in size from left to right.

Each of the nine primary segments is subdivided into ten spaces using **secondary graduations** and, space permitting, each of those spaces is further divided into ten spaces using **tertiary graduations**, as shown in the box to the right.

When reading a logarithmic scale, the number indicated by an arrow is determined by which primary, secondary, and tertiary (if applicable) segment the arrow lies. For example, the number 1.45 indicates that the

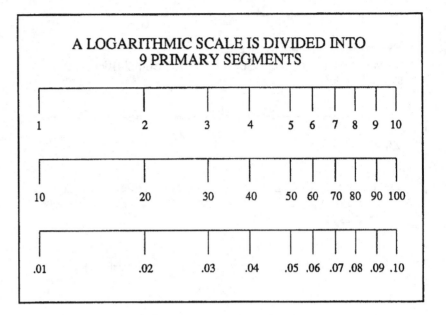

A LOGARITHMIC SCALE IS DIVIDED INTO 9 PRIMARY SEGMENTS

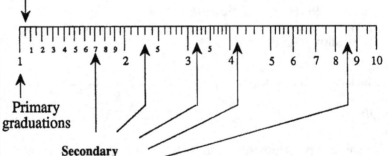

EACH PRIMARY SEGMENT HAS 10 SECONDARY SEGMENTS

EACH SECONDARY SEGMENT HAS 10 TERTIARY SEGMENTS

Tertiary graduations divide each secondary space into ten parts. (Sometimes all nine tick marks are shown, or every other tick mark. More frequently, however, only one tick mark is shown, the 5 value.)

Primary graduations

Secondary graduations
divide each primary space into 10 parts. (Sometimes all nine tick marks are shown, as for primary segments 1, 2 and 3; sometimes only every other tick mark is shown, as for primary segments 4, 5 and 6; and sometimes only one tick mark is shown, the 5 mark as shown for primary segments 7, 8 and 9.)

---

* A detailed discussion of logarithms is beyond the scope of this text.

** Refer to Chapter 13 for a review of powers.

Example 10: (Reading Scales)

❑ Given the logarithmic scale shown below, give the value corresponding to each arrow.

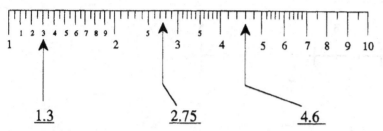

1.3

The arrow falls within the 1 primary segment, so the first digit is 1. The arrow falls on the third secondary unit, so the second digit is 3.

2.75

The arrow falls within the 2 primary segment, so the first digit must be a 2. It falls within the 7 secondary segment, so the second digit is a 7. Since the arrow is about halfway between the 7 and 8 secondary units, the tertiary reading is a 5.

4.6

The arrow falls within the 4 primary segment, so the first digit is a 4. There are only four tick marks dividing the primary segment into 10 parts. Thus each tick mark has a value of 2. The second digit is a 6.

Example 11: (Reading Scales)

❑ What is the value corresponding to each arrow shown on the logarithmic scale below?

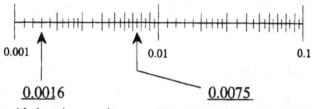

0.0016

Begin with the primary unit 0.001, then add any secondary and tertiary digits to the right of the primary unit.

0.0016

The secondary segment is divided into 10 parts, so each tick mark represents a value of 2; therefore the secondary digit is a 6.

0.0075

The primary segment is 0.007. To this unit add the secondary unit. From 0.007 to 0.008 there is only one tick mark, representing the value of 5. No tertiary reading is possible on this scale.

---

point lies within the first primary segment, within the fourth secondary segment and at the 5 tertiary mark:

Many graphs that use logarithmic scales include only primary scales. Others include only primary and secondary scales. **Readings that fall between tick marks must be visually estimated.** Examples 10 and 11 illustrate readings made from a logarithmic scale.

PLACING THE DECIMAL POINT

The best way to place the decimal point for a logarithmic scale reading is to note the location of the decimal point in the primary unit **to the left** of the given or desired point. In Example 11, the primary unit to the left of the first arrow is a 1 (1.0). Therefore the decimal point in the scale reading is placed as 1.3.

WHEN THE SCALE INCLUDES DECIMAL FRACTIONS*

Reading scales which include decimal fractions (decimal numbers less than one) tends to be more difficult than reading scales with whole numbers. To make it easier, remember that digits read from secondary and tertiary graduations are placed **to the right of the primary unit reading.** For example:

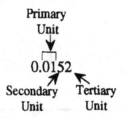

---

* For a review of decimal fractions, refer to Chapter 4.

PRACTICE PROBLEMS 12.1

❑ In the following problems, give the mixed number or fraction (in reduced form) for each point on the arithmetic scales given below.

1.

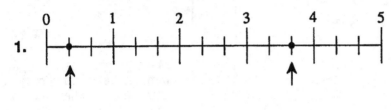

2.

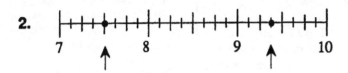

3.

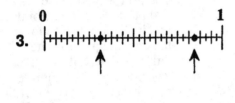

❑ Give the value of each point on the logarithmic scales given below.

4.

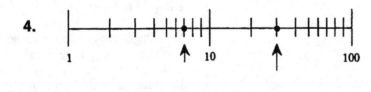

5.

12.2 TYPES OF GRAPHS

SUMMARY

The line graph, bar graph and circle graph are three types of graphs commonly used to present and analyze data:

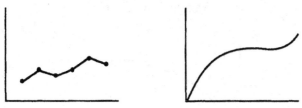

Broken-Line Graph   Smooth-Line Graph

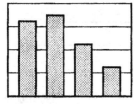

Bar Graph

Divided Bar Graph

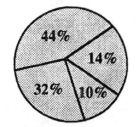

Circle Graph

Another type of graph, used as a shortcut in making various calculations, is called a nomograph. This type of graph utilizes three or more different scales.

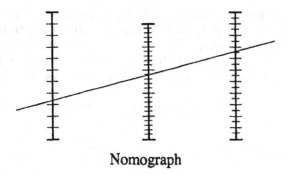

Nomograph

LINE GRAPHS

Of the three types of graphs—line graphs, bar graphs, and circle graphs—line graphs are by far the most commonly used graphs in the water and wastewater field. They are relatively simple to construct and can display several sets of data on the same graph. Line graphs are often used to depict changes in data over a given time period.

The two types of line graphs are the **broken-line graph** and the **smooth-line graph.** Generally, treatment system data presented in a line graph will result in a broken-line graph. Curved-line graphs are derived from complex mathematical equations.

READING LINE GRAPHS

The primary skill in reading line graphs is that of reading scales, discussed in the previous section. Each graph includes a horizontal scale, measured along the horizontal axis, and a vertical scale, measured along the vertical axis. The mathematical terms used for these scales are **abscissa** (horizontal scale) and **ordinate** (vertical scale). *Hint:* One way to remember which term applies to which scale is to note the first letters of each word. The two words with first letters closest to "a" go together, (abscissa and horizontal); and the two words closest to "z" go together (ordinate and vertical).

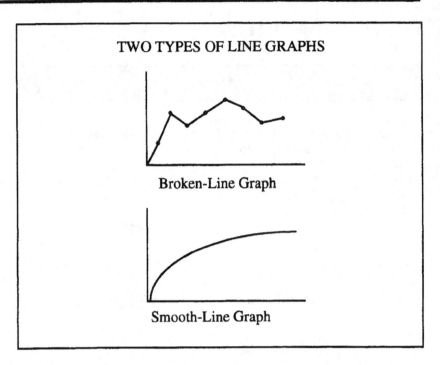

TWO TYPES OF LINE GRAPHS

Broken-Line Graph

Smooth-Line Graph

Example 1: (Types of Graphs)

❑ Given the graph below, when the organic loading is 100 lbs BOD/day/1000 cu ft, what is the approximate value?

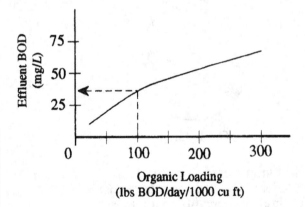

Organic Loading
(lbs BOD/day/1000 cu ft)

From the given value on the horizontal scale (100 lbs BOD/day/1000 cu ft), move vertically up to the graphed line. Then, from the point of intersection, move horizontally to the left scale and read the approximate value. The value is approximately halfway between the 25 and 50 marks: 37.5 mg/*L*.

Example 2: (Types of Graphs)

❏ Construct a line graph given the following consecutive average daily flows: 1.6 MGD; 1.4 MGD; 1.2 MGD; 1.3 MGD; 1.5 MGD; 1.2 MGD; 1.1 MGD.

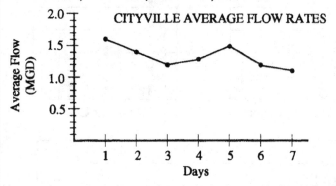

Example 3: (Types of Graphs)

❏ Using the log-log graph below, determine the flow rate (in cfs) at 0.5 feet of head on the weir.

90° V-NOTCH WEIR (Formula $Q = 2.5\,H^{5/2}$)

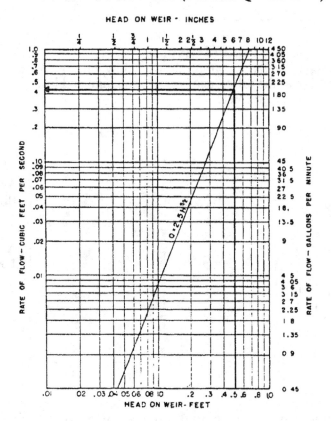

(Source: New York Manual of Instruction for Water Treatment Plant Operators.)

From 0.5 ft of head on the bottom scale, move vertically upward to the point of intersection, then move horizontally to the left scale. Flow = 0.45 cfs.

To read a line graph:

1. Find the known value on the horizontal scale.

2. From that point, move vertically up until you intersect the line graph.

3. Then, from the point of intersection, move horizotally toward the left (or right) scale.

4. Read the value on the vertical scale.

This same basic process can be used if the known value is on the vertical scale—from the vertical scale to the intersection point and down to the horizontal scale.

CONSTRUCTING LINE GRAPHS

When constructing line graphs follow these steps:

1. Draw the horizontal and vertical axes and mark off the scales. Time data (days, months, years) are generally placed on the horizontal axis.

   The scales for these graphs are normally arithmetic scales. This means that the tick marks on each scale must be equally-spaced. (One scale can, however, have a different scale than the other.)

2. Plot each data point on the graph.

3. Connect the dots.

Example 2 illustrates the construction of a simple line graph.

SEMI-LOG OR LOG-LOG GRAPHS

Graphs that have logarithmic scales on only one axis, with an arithmetic scale on the other axis, are called **semi-log graphs**. Graphs that have logarithmic scales on both axes are called **log-log graphs**. Example 3 illustrates a log-log graph.

BAR GRAPHS

Bar graphs are often used for comparing data. The bars may be arranged vertically, as shown to the right, or horizontally, if desired.

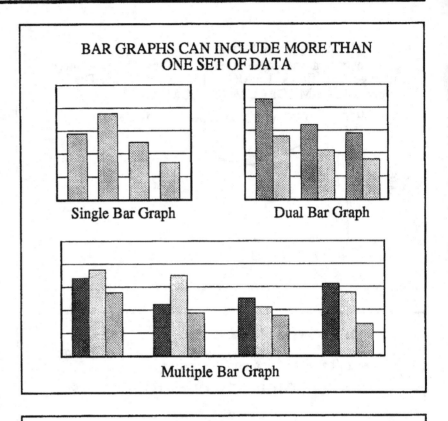

BAR GRAPHS CAN INCLUDE MORE THAN ONE SET OF DATA

Single Bar Graph   Dual Bar Graph

Multiple Bar Graph

READING BAR GRAPHS

Reading a bar graph is a simple matter. From the top of any bar, move horizontally to the scale and read the value. This may be done for any bar on the graph.

More importantly, bar graphs aid in the comparison of data and illustrate the overall progression or trend in the data.

Example 4: (Types of Graphs)
❑ Based on the bar graph shown below, during which month was the difference between the high and low temperature greatest?

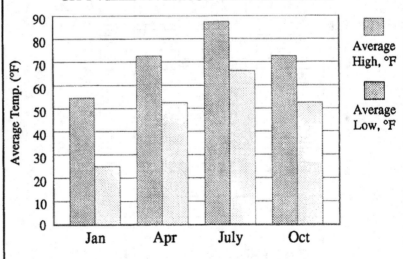

CITYVILLE AVERAGE TEMPERATURES

Average High, °F

Average Low, °F

For this question, you must determine which month has the greatest difference in high and low bar heights. The answer is **January**.

Example 5: (Types of Graphs)
❑ Given the following average daily flow data, construct a bar graph: M—1.6 MGD; T—1.4 MGD; W—1.2 MGD; Th—1.3 MGD; F—1.5 MGD; Sa—1.2 MGD; Su—1.1 MGD

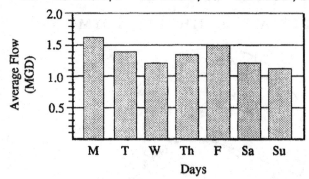

First, determine the scale size. You may have to experiment with various sized scales to ensure that the bar lengths are easily read and yet seem balanced. Remember, time data is normally shown along the horizontal scale.

Next, label the scales and complete the bars.

Example 6: (Types of Graphs)
❑ A small treatment system staff is comprised of 23% administrative, 38% operations, 12% maintenance, 19% collections, and 8% engineering. Present this data as a divided bar graph.

8% Engineering
19% Collections
12% Maintenance
38% Operations
23% Admin.

The scale length must first be selected. (3 in. is selected due to space considerations.) Then the length of each part is calculated:**

(0.23) (3 in.) = 0.7 in. Admin.

(0.38) (3 in.) = 1.1 in. Oper.

(0.12) (3 in.) = 0.4 in. Maint.

(0.19) (3 in.) = 0.6 in. Coll.

(0.08) (3 in.) = 0.2 in. Engr.

The length of each part is then drawn on the bar graph and labeled.

CONSTRUCTING BAR GRAPHS

Selection of appropriately sized scales is the most important aspect in constructing any graph. In the case of a bar graph, if the scale is too expanded, the bars will appear excessively long. If the scale is too compressed, the bars will appear squatty with the differences in bar heights hard to distinguish. A break in the scale may be necessary so that bars are not too long. A break in the scale is denoted as:

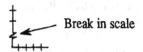

Sometimes data to be graphed are single data points, as shown in Example 5. In other cases the data points represent average* values from a much larger data base.

Be sure to label your scales and title the graph. Without these the information is useless.

DIVIDED BAR GRAPH

A **divided bar graph** is similar to a circle graph, described on the next pages. This type graph may be used when the data involves percentages of a whole.

To construct such a graph, establish the length of the bar. Again, you may need to experiment to find a length that is aesthetically pleasing as well as making the data easy to read.

Once the total bar length is determined (e.g., 3 inches), multiply the percent represented by each part by the total length. This calculation will determine the lengths of each part within the divided bar. Example 6 illustrates this procedure.

---

* Refer to Chapter 6 for a discussion of averages.
** Refer to Chapter 5 for a review of percent calculations.

CIRCLE GRAPHS

A circle graph is often used to display relative proportions or percents of various data compared to the whole. **The only difference between this type graph and a divided bar graph is format—** here the data is presented as a circle, not a rectangle. Because the wedge-shaped parts of a circle graph resemble the shape of pie pieces, this type of graph is often referred to as a **pie chart.**

The construction of a circle graph may be more difficult, because it includes drawing angles, however its format may be more desirable than a divided bar graph. It is strictly a matter of personal preference.

Example 7: (Types of Graphs)

❑ Based on the circle graph given below, which length of employment characterizes the greatest number of staff?

DISTRICT STAFF LENGTH OF EMPLOYMENT

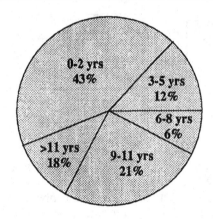

The length of employment for the greatest number of staff is 0-2 yrs.

Example 8: (Types of Graphs)

❑ On the average, about 4.2 trillion gallons of precipitation fall on the United States every day. The circle graph below indicates the destination of that water. About what percent of the precipitation is used directly by people?

U.S. WATER BUDGET

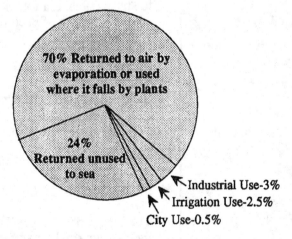

The use by people includes industrial, irrigation, and city use: 3% + 2.5% + 0.5% = 6%

MEASURING ANGLES WITHIN A CIRCLE

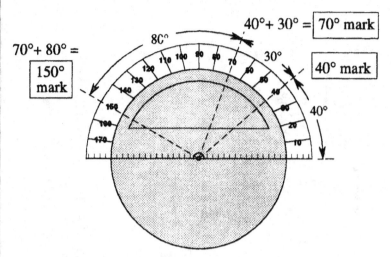

$70° + 80° = \boxed{150° \text{ mark}}$

$40° + 30° = \boxed{70° \text{ mark}}$

$\boxed{40° \text{ mark}}$

To measure several different angles from the same protractor setting, add the degrees of the new angle to the previous angle measurement. When the measure reaches 180°, the protractor must be moved. It can be flipped 180° and the process continued for the full 360°.

CONSTRUCTING CIRCLE GRAPHS

To construct a circle graph you must first determine how much of the circle is to be represented by each part. This is done by multiplying 360° by the percent or ratio represented by that part. For example, if one type of data represents 20% then:

$$360° \times \frac{20}{100} = 72°$$

Thus, 72° of the circle is to be represented by that data.

Once you have calculated the degrees of the circle represented by each part, then measure these angles in the circle.

1. Draw a line from the center of the circle to the edge (a radius).

2. Place a protractor such that the flat edge of the protractor is on the radius, with center points matching.

3. From the right side of the protractor, read and mark the first desired degrees (e.g., 40°).

4. Marking the second and subsequent desired degrees may be done in a couple of ways. The protractor may be placed on each new radius drawn and a new angle constructed from that point. The disadvantage of this method is that each time the protractor is lifted and placed again, it is likely that a few degrees of error are introduced; therefore, it is advisable to move the protractor as little as possible. To measure a second angle without moving the protractor, add the degrees of the first and second angles and measure that combined angle on the protractor. The box at the top of the page illustrates this process.

Example 8: (Types of Graphs)
❑ Use the same data as given in Example 6 to construct a circle graph.

DISTRICT STAFF

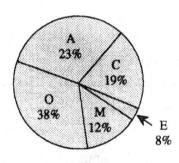

Key:
A = Administration
O = Operations
M = Maintenance
C = Collection
E = Engineering

First, determine the part of the circle that is to be represented by each part:

$(360°) (0.23) = 83°$ Admin.

$(360°) (0.38) = 137°$ Oper.

$(360°) (0.12) = 43°$ Maint.

$(360°) (0.19) = 68°$ Coll.

$(360°) (0.08) = 29°$ Engr.

Then mark these degrees on the circle graph using a protractor and label each part. If space permits, place labels within the circle. If not, a key may be used, as shown above.

* Refer to Chapter 7 for a review of proportions.

NOMOGRAPHS

Line graphs, bar graphs, and circle graphs are three types of graphs commonly used for data presentation and analysis. Another type of graph, a nomograph, is used as a shortcut to mathematical calculations.

At least three scales are used to determine the desired answer on the nomograph. Values are known along two of the three scales. A straight line is then drawn from the two known values to the third scale, indicating the desired value.

There are many other types of nomographs than the one shown on this page. Some include curved scales, others include "pivot points" and multiple scales. The type of nomograph depends on the type of mathematical equation it represents. The nomographs you will encounter in water and wastewater treatment will likely be very similar to that shown on this page.

It should be emphasized that nomographs give approximate answers. If a precise answer is desired, mathematical calculation is required.

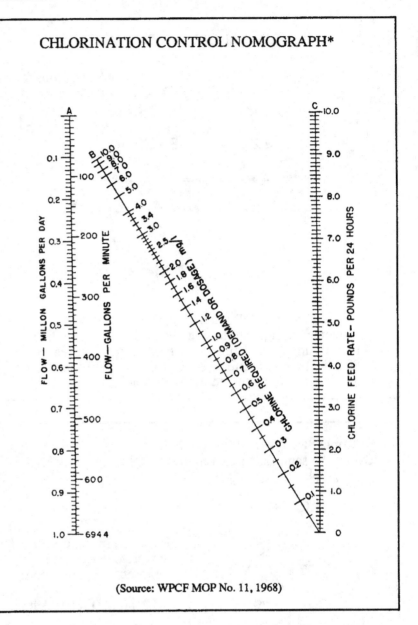

CHLORINATION CONTROL NOMOGRAPH*

(Source: WPCF MOP No. 11, 1968)

Example 10: (Types of Graphs)
❏ Using the nomograph above, if the flow rate is 0.8 MGD and the required chlorine dosage is 0.7 mg/L, how many lbs/day chlorine feed rate will be required?

Lay a straightedge at the 0.8 mark on Scale A (left side of the scale) and rotate the straightedge until it also aligns with the 0.7 mark on Scale B.

Draw a line along the straightedge until it intersects with Scale C. The value at that point should be 4.7 lbs/day.

---

* The values obtained using this nomograph can be verified using the mg/*L* to lbs/day calculation. Refer to Chapter 8, Conversions, for a review of these calculations.

Example 11: (Types of Graphs)
❑ What would the chlorine dosage be, in mg/L, if the flow rate is 500 gpm and the chlorine feed rate is set at 7.2 lbs/day. Use the nomograph to the left.

First lay a straightedge on Scale A (right side of scale) at 500 gpm. Rotate the straightedge until it also aligns with 7.2 on Scale C.

Draw a line along the straightedge and read the value where the line intersects with the middle scale (Scale B):

$$\boxed{1.2 \text{ mg/}L}$$

Example 12: (Types of Graphs)
❑ The required chlorine dosage rate is 4 mg/L. If the flow to be treated is 3.5 MGD, what is the required chemical feed rate in lbs/day?

First, align the straightedge at 3.5 on Scale A. Since the scale only includes the values between 0 and 1 MGD, the flow rate value will have to be adjusted temporarily. If the decimal point is moved **one place to the left**, a value is obtained that is *on* the scale:

$$3.5 = 0.35 \text{ MGD}$$

When the straightedge is placed at 0.35 on Scale A and 4 mg/L on Scale B, the line drawn along the straightedge falls above Scale C. Therefore, the mg/L value must also be temporarily adjusted. Move the decimal point **one place to the left**:

$$4 = 0.4 \text{ mg/}L$$

Align the new values (0.35 on Scale A and 0.4 on Scale B) and read the answer on Scale C:

$$1.15 \text{ lbs/day}$$

Now readjust the decimal point in the answer. A total of **two moves to the right** is required:

$$1.15 = \boxed{115 \text{ lbs/day}}$$

---

* Refer to Chapter 13 for a review of powers of 10.

ANY TWO SCALES MAY BE USED

In Example 10, the left and center scales were used to determine a value on the right scale. In fact, **values from any two scales can be used to determine a value on the third scale:**

1. Align the straightedge at a given value on one of the scales.

2. Rotate the straightedge so that it is also aligned with a given value on either of the remaining two scales.

3. Draw a line passing through both given values until it intersects the third scale.

4. Read the value at the point of intersection with the third scale.

WHEN VALUES EXCEED THE SCALE

This nomograph may also be used for values that exceed the scales. For example, although 3 MGD exceeds the values given on Scale A, it can be temporarily adjusted by a factor of ten.* Simply move the decimal point one place to the left:

$$3.0 = 0.3$$

This value (0.3 MGD) *is* given on Scale A. Once an answer has been obtained using the adjusted factor, be sure to adjust the decimal point back to its correct place. (In this example, move the decimal point in the final answer one place to the right.)

Occasionally, both values may require a temporary decimal point adjustment so that the nomograph may be used, as illustrated in Example 12.

PRACTICE PROBLEMS 12.2: Types of Graphs

1. According to the graph shown below, what was the approximate chlorine demand (mg/*L*) when 2.0 mg/*L* of chlorine was added?

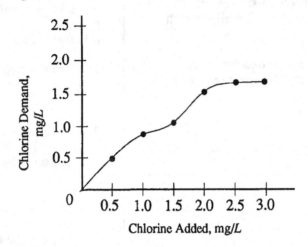

ANS_____

2. Graph the following data as a line graph.

	Jar Test 1	Jar Test 2	Jar Test 3	Jar Test 4	Jar Test 5	Jar Test 6
Catonic Polymer, mg/*L*	0.5	1.0	1.5	2.0	2.5	3.0
Settled Water Turbidity	0.75	0.45	0.30	0.20	0.35	0.70

PRACTICE PROBLEMS—CONT'D

3. Using the graph shown below, if the head on the weir is 6 inches, what is the cubic feet per second flow rate through the 90° V-notch weir?

90° V-NOTCH WEIR (Formula $Q = 2.5\,H^{5/2}$)

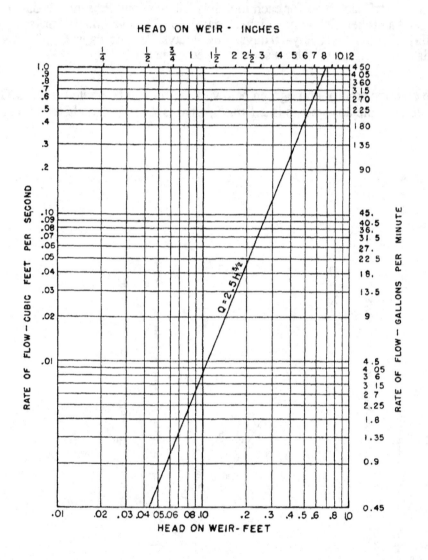

ANS_____

PRACTICE PROBLEMS—CONT'D

4. A treatment system staff is comprised of 18% administrative, 36% operations, 15% maintenance, 22% collections, 9% engineering. Present this data as both a divided bar graph and a circle graph. (Use a separate sheet of paper.)

5. A summary of staff sick days for each month is listed below. Present this data as both a bar graph and a broken-line graph. (Jan-20 days; Feb-28 days; Mar-10 days; Apr-12 days; May-14 days; June-8 days; July-10 days; Aug-9 days; Sept-16 days; Oct-21 days; Nov-18 days; Dec-25 days) (Use a separate sheet of paper.)

6. Using the chlorine control nomograph below, if the flow to be treated is 0.5 MGD, and the required chlorine dosage rate is 2.1 mg/L, what is the required chemical feed rate, in lbs/day?

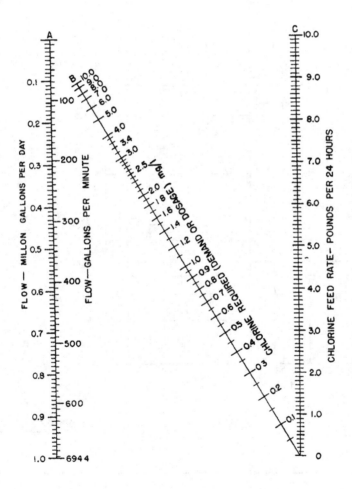

ANS_____

13 *Powers, Roots, and Scientific Notation*

13.1 Positive Exponents

Number Correct

❑ Write the following numbers in expanded form as factors.

1. $7^2 =$ _____

3. $5^0 =$ _____

2. $12^3 =$ _____

4. $3^2 \cdot 2^3 =$ _____

❑ Write the following numbers using exponential notation.

5. $(4)(4) =$ _____

7. $(x)(x)(x) =$ _____

6. $(16)(16)(3)(3)(3) =$ _____

8. $(0.785)(90)(90) =$ _____

❑ Complete the following problems.

9. $(0.785)(75^2) =$ _____

10. $2^3 \cdot 3^3 =$ _____

13.2 Negative Exponents

Number Correct

❑ Write the following numbers in expanded factor form.

1. $70^{-2} =$ _____

3. $4^{-3} \cdot 3^0 =$ _____

2. $15^{-1} =$ _____

4. $50^2 \cdot 2^{-3} =$ _____

❑ Write the following numbers using exponential notation. (The answers should include only factors with positive or negative exponents, as shown in problems 1-4.)

5. $\dfrac{1}{(25)(25)} =$ _____

7. $\dfrac{(0.785)(60)(60)}{(20)(20)} =$ _____

6. $\dfrac{1}{(3)(3)(3)(3)} =$ _____

8. $\dfrac{(28)(28)}{50} =$ _____

❑ Complete the following problems.

9. $(2^{-5})(6^2) =$ _____

10. $(3^2)(4^{-3}) =$ _____

13.3 Fractional Exponents

Number Correct

❑ Write the following numbers in radical form.

1. $225^{1/2} =$ _____

3. $2.5^{5/2} =$ _____

2. $40^{1/3} =$ _____

4. $4^{3/2} =$ _____

13.3 Fractional Exponents—Cont'd

❏ Write the following numbers in exponential form.

5. $\sqrt{1600}$ = _____

7. $8\sqrt{6358.5}$ = _____

6. $\sqrt[3]{27}$ = _____

8. $\sqrt{(45)^2}$ = _____

❏ Complete the following problems.

9. $4900^{1/2}$ _____

10. $\sqrt{(9)^3}$ = _____

13.4 Multiplying and Dividing Powers

❏ Simplify the following exponential expressions using the law of exponents.

1. $2^3 \cdot 2^2$ = _____

4. $\dfrac{(15^2)(2^3)(7)}{(7^3)(2)}$ = _____

2. $\dfrac{5^2 \cdot 3^3}{5^3}$ = _____

5. $\dfrac{(3)(4^2)(x^3)}{(x)}$ = _____

3. $\dfrac{7x^2}{x}$ = _____

13.5 Powers of Ten

❑ Complete the following multiplication and division problems using only decimal point changes.

1. 45×100 = _____

2. $\dfrac{67.5}{100}$ = _____

3. $(0.003)(10{,}000)$ = _____

4. $\dfrac{12.85}{10{,}000}$ = _____

5. $\dfrac{(27.46)(10)}{100}$ = _____

13.6 Scientific Notation

❑ Write the following numbers as decimal numbers.

1. 1.62×10^{3} = _____

2. 4.79×10^{-4} = _____

3. 9.154×10^{-2} = _____

4. 6.44×10^{6} = _____

❑ Write the following numbers in scientific notation.

5. $4{,}150{,}000$ = _____

6. $326{,}000$ = _____

7. 0.00062 = _____

8. 2500 = _____

13.6 Scientific Notation—Cont'd

❏ Complete the following problems.

9. $(2.92 \times 10^6)(1.45 \times 10^3) = $ _____

10. $(6.67 \times 10^{-2})(7.14 \times 10^3) = $ _____

NOTES:

13.1 POSITIVE EXPONENTS

SUMMARY

1. An **exponent** indicates how many times a number is to be multiplied together.

2. The **base** is the number that is being multiplied.

$$7^4 = (7)(7)(7)(7)$$

where the 4 is the exponent and the 7 is the base.

3. The entire expression, such as 7^4, is called a **power**. It is read "seven to the fourth power".

4. These same considerations apply to letters as well. For example:

$$x^2 = (x)(x) \quad \text{or} \quad x^3 = (x)(x)(x)$$

5. Any number or letter that does not have an exponent is considered to have an exponent of one.
   Thus $10 = 10^1$ or $x = x.^1$

The use of exponents is a "shorthand" method of writing multiplication of the same factor two or more times. The factor or number being multiplied is called the **base**. The number of times it is used as a factor is called the exponent. For example:

$$\text{base} \rightarrow 2^4 \leftarrow \text{exponent}$$

The base and exponent together, such as 2^4, is called a **power**. Two to the fourth power means that two is multiplied together four times:

$$2^4 = (2)(2)(2)(2)$$

When a number is written with an exponent, it is said to be in **exponential notation**. When all the facors are written, it is in expanded factor form. Letters can also be written using exponential notation. Thus, $x^2 = (x)(x)$, etc.

Examples 1-4 illustrate calculations involving positive exponents of one or greater.

Example 1: (Positive Exponents)
❑ Write the following numbers in expanded form: 6^2, 20^3, 3^5, and x^3.

$$6^2 = \boxed{(6)(6)}$$

$$20^3 = \boxed{(20)(20)(20)}$$

$$3^5 = \boxed{(3)(3)(3)(3)(3)}$$

$$x^3 = \boxed{(x)(x)(x)}$$

Example 2: (Positive Exponents)
❑ Write the factors shown below using exponential notation.

$$(4)(4)(4) = \boxed{4^3}$$

$$(2)(2)(2)(2) = \boxed{2^4}$$

$$(7)(7)(7)(7)(7)(7) = \boxed{7^6}$$

$$(x)(x) = \boxed{x^2}$$

Example 3: (Positive Exponents)

❑ Write the following factors using exponential notation.

$$(8)(8)(8)(7)(7) = \boxed{8^3 \, 7^2}$$

$$(6)(x)(x)(x) = \boxed{6x^3}$$

$$(20)(20)(3)(3)(4) = \boxed{20^2 3^2 4}$$

Example 4: (Positive Exponents)

❑ Complete the following problems.

$$6^2 3^3 = (6)(6)(3)(3)(3)$$

$$= \boxed{972}$$

$$4^3 2^5 = (4)(4)(4)(2)(2)(2)(2)(2)$$

$$= \boxed{2048}$$

EXPONENTS OF ONE

When a number or letter has no exponent, it is considered to have an exponent of one:

$$2 = 2^1$$

Because an exponent indicates how many times a factor is to be used, an exponent of one does not have much meaning in terms of multiplication. (At least two numbers are needed for a multiplication problem.) However, when multiplying and dividing powers, it is an important concept.

(*Note:* 2^0 does not equal 2 x 1, but simply 1.)

GROUP ONLY FACTORS WITH THE SAME BASE

When there is a string of factors, they may be written in exponential form by **grouping like factors**. For example:

$$(3)(3)(2)(2)(2) = 3^2 2^3$$

A dot is sometimes used to indicate multiplication:

$$3^2 2^3 = 3^2 \cdot 2^3$$

PRACTICE PROBLEMS 13.1: Positive Exponents

❏ Write the following terms in expanded factor form.

1. $2^4 6^2 = $ _____

3. $9^2 \cdot 4^6 = $ _____

2. $5x^3 = $ _____

4. $2 \cdot 2^2 \cdot 3^3 = $ _____

❏ Write the following factors using exponential form.

5. $(2)(3)(3) = $ _____

7. $(8)(x)(x) = $ _____

6. $(3)(3)(3) = $ _____

8. $(2)(2)(2)(2)(2) = $ _____

❏ Complete the following calculations.

9. $6^3 4^2 = $ _____

10. $9^2 2^3 x^2 = $ _____

13.2 NEGATIVE EXPONENTS

SUMMARY

1. A factor with a **negative exponent** can be inverted and written with a positive exponent, as follows:

$$3^{-2} = \frac{1}{3^2}$$

2. Any number that has an **exponent of zero** is equal to one. For example:

$$7^0 = 1$$

$$4^0 = 1$$

$$x^0 = 1$$

3. **When a power is moved** from the numerator of a fraction to the denominator, or vice versa, the sign of the exponent must be changed. For example:

$$\frac{2^2 \cdot 3^{-2}}{6} = \frac{2^2}{6 \cdot 3^2}$$

Although negative exponents often present no difficulty in making mathematical calculations, it is sometimes advantageous to convert the power from a negative exponent to a positive exponent. This may be accomplished by **moving the power**.

For terms that include only multiplication and division, powers may be moved from the numerator to the denominator provided the sign of the exponent is changed, as shown in the box at the top of this page.

NEGATIVE EXPONENTS CAN BE WRITTEN AS POSITIVE EXPONENTS

When moving a power from the numerator* to the denominator (or vice versa), change the sign of the power.

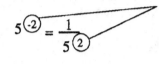

The sign of the power is changed as that power is moved from the numerator to the denominator.

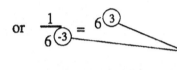

The sign of the power is changed as that power is moved from the denominator to the numerator.

Example 1: (Negative Exponents)
❑ Write the following terms in exponential notation using only positive exponents:

$$2^{-2}\,3^{\,4}; \quad 5^2\,3^{-3}\,2^{-2}; \quad \frac{3^2 5^2}{2^{-3}}$$

Each negative exponent must be moved in order for the sign to be changed to positive:

$$2^{-2}\,3^{\,4} = \boxed{\dfrac{3^4}{2^2}}$$

$$5^2\,3^{-3}\,2^{-2} = \boxed{\dfrac{5^2}{3^3 2^2}}$$

$$\frac{3^2 5^2}{2^{-3}} = \boxed{3^2\,5^2\,2^3}$$

---

* A whole number is always considered to be over one; $3 = 3/1$ and $17 = 17/1$. Thus, a power written as 6^2 is considered to be in the numerator: $6^2/1$.

Example 2: (Negative Exponents)
❏ Write the following terms in expanded form using factors only (no exponents): $10x^{-2}$; $2^5 \cdot 4^{-2}$; $\dfrac{6^0 x^3}{x^{-2}}$

$$10x^{-2} = \frac{10}{x^2} = \boxed{\frac{10}{(x)(x)}}$$

$$2^5\, 4^{-2} = \frac{2^5}{4^2} = \boxed{\frac{(2)(2)(2)(2)(2)}{(4)(4)}}$$

$$\frac{6^0 x^3}{x^{-2}} = 6^0\, x^3 x^2 = \boxed{(1)(x)(x)(x)(x)(x)}$$

ZERO EXPONENTS

When a number has a zero exponent, it is always equal to one:

$$5^0 = 1 \quad \text{or} \quad x^0 = 1$$

Example 3: (Negative Exponents)
❏ Complete the following calculations:
$(0.785)(60^2)(20)$; and $(3^2)(2^{-3})(6^2)$

$$(0.785)(60^2)(20) = (0.785)(60)(60)(20)$$

$$= \boxed{56{,}520}$$

$$3^2\, 2^{-3}\, 6^2 = \frac{(3)(3)(\overset{3}{\cancel{6}})(\overset{3}{\cancel{6}})}{(2)(\cancel{2})(2)}$$

$$= \boxed{40.5}$$

DIVIDE OUT FACTORS WHENEVER POSSIBLE

As with other calculations, divide out common factors whenever possible. This makes the calculation less cumbersome.

PRACTICE PROBLEMS 13.2 Negative Exponents

❑ Write the following terms in exponential notation using only positive exponents.

1. $6x^{-2} =$ _____

3. $(8^2)(5^{-3}) =$ _____

2. $\dfrac{3^{-2}2^4}{7^3} =$ _____

4. $\dfrac{9^{-3}x^2}{2^{-2}} =$ _____

❑ Write the following terms in expanded form using factors only (no exponents).

5. $\dfrac{2^3 3^{-2}}{5} =$ _____

7. $3^4 \cdot 2^{-3} \cdot 5^0 =$ _____

6. $\dfrac{9x^2}{2^{-2}} =$ _____

8. $10^{-2} 6^2 =$ _____

❑ Complete the calculations given below.

9. $5^2 \cdot 10^{-2} =$ _____

10. $\dfrac{(15^2)(2^{-3})}{5} =$ _____

13.3 FRACTIONAL EXPONENTS

SUMMARY

1. A **root** is a number which, when multiplied together two (or more) times, equals the original number.

   A **square root** is a number which, when multiplied together **twice**, equals the original number. For example, the square root of 64 is 8, since 8 x 8 = 64.

   A **cube root** is a number which, when multiplied together **three times**, equals the original number. For example, the cube root of 8 is 2, since 2 x 2 x 2 = 8.

2. A **fractional exponent indicates a root is to be taken.** The denominator of a fracitonal exponent determines which root is to be taken.

3. **A fractional exponent of 1/2 indicates that a square root** is to be taken. A square root may also be written as a radical:

$$x^{1/2} = \sqrt{x}$$

4. **A fractional exponent of 1/3 indicates that a cube root** is to be taken. A cube root may also be written as a radical:

$$x^{1/3} = \sqrt[3]{x}$$

5. The **numerator of the fractional exponent** is the power of the base. For example:

$$8^{2/3}$$

The root to be taken

The power of 8

The base

This may be written using a radical as:

$$\sqrt[3]{8^2}$$

THE ROOT OF A NUMBER

A **root** is a number which, when multiplied together a given number of times, equals the original number. The type of root (square root, cube root, fourth root, etc.) determines how many times a number must be multiplied to equal the original number.

The most common root you will encounter in water and wastewater calculations is square root. A **square root** is a number, which when multiplied together **two times**, equals the original number. For example, the square root of 49 is 7, since 7 x 7 = 49; the square root of 144 is 12, since 12 x 12 = 144.

A **cube root** is a number, which when multiplied together **three times**, equals the original number. As an example, the cube root of 8 is 2, since 2 x 2 x 2 = 8.

Taking a square root, cube root, or any other root is denoted in either of two ways:

1. By using a radical, or

2. By using a fractional exponents.

FRACTIONAL EXPONENTS INDICATE BOTH POWERS AND ROOTS OF A NUMBER

$120^{3/2}$

The **numerator** indicates the power of the number

The **denominator** indicates the root of the number

This can be written in radical form:

$\sqrt{120^3}$

The radical indicates the root of the number

The exponent indicates the power of the number

Example 1: (Fractional Exponents)

❑ Express the following numbers using radicals: $4^{1/2}$, $64^{1/3}$, $90^{2/3}$, and $1.4^{5/2}$

$$4^{1/2} = \sqrt{4}$$

$$64^{1/3} = \sqrt[3]{64}$$

$$90^{2/3} = \sqrt[3]{90^2}$$

$$1.4^{5/2} = \sqrt{1.4^5}$$

Example 2: (Fractional Exponents)
❑ Express the following numbers using fractional exponents:
$$\sqrt{5^3}\;;\;\sqrt[3]{x^2}\;;\;\sqrt[5]{150}$$

$$\sqrt{5^3} = \boxed{5^{3/2}}$$

$$\sqrt[3]{x^2} = \boxed{x^{2/3}}$$

$$\sqrt[5]{150} = \boxed{150^{1/5}}$$

Example 3: (Fractional Exponents)
❑ Complete the following calculations:
$$484^{1/2}\;;\;\sqrt[3]{27}\;;\;4^{3/2}$$

$$484^{1/2} = \sqrt{484} = \boxed{22}$$

$$\sqrt[3]{27} = \boxed{3}$$

$$4^{3/2} = \sqrt{4^3} = \sqrt{64}$$

$$= \boxed{8}$$

RADICALS AND FRACTIONAL EXPONENTS

A **radical** is a mathermatical symbol ($\sqrt{}$) that indicates that a root of a number is desired. For example, the square root of 100 may be written as:

$$\sqrt{100}$$

For roots, **other than square roots**, the desired root must be included with the radical. Thus, the cube root of 8 is written as:

$$\sqrt[3]{8}$$

Fractional exponents may also be used to denote roots. The denominator of the fractional exponent indicates the desired root. Using the two illustrations above, the square root of 100 and cube root of 8 would be written as follows, using fracitonal exponents:

$$100^{1/2} \text{ and } 8^{1/3}$$

The numerator of a fractional exponent indicates the power of the number; and as such, when written in radical form, the power is placed under the radical with the number. To illustrate:

$$2.5^{5/2} = \sqrt{2.5^5}$$

Note that the denominator of the fractional exponent (2) denotes a square root is desired, and the numerator (5) indicates that 2.5 is to the fifth power.

PRACTICE PROBLEMS 13.3: Fractional Exponents

❏ Express the following numbers using radicals.

1. $20^{1/2} = $ _____

3. $50^{2/3} = $ _____

2. $1728^{1/3} = $ _____

4. $2.8^{5/2} = $ _____

❏ Express the following numbers using fractional exponents.

5. $\sqrt{30^3} = $ _____

7. $\sqrt[4]{7^3} = $ _____

6. $\sqrt[3]{25^2} = $ _____

❏ Complete the calculations shown below.

8. $25^{3/2} = $ _____

10. $3^2 \cdot 4^{1/2} = $ _____

9. $169^{1/2} \cdot 144^{1/2} = $ _____

13.4 MULTIPLYING AND DIVIDING POWERS

SUMMARY

1. When **multiplying powers** with the same base, simply add exponents:

$$x^2 \cdot x^3 = x^{2+3} = x^5$$

2. When **dividing powers** with the same base, subtract the power of the denominator from the power of the numerator:

$$\frac{x^5}{x^3} = x^{5-3} = x^2$$

MULTIPLYING POWERS

Exponents indicate the number of times a factor (or base) is to be multiplied. When powers of the same base are multiplied, the number of factors represented by the exponent of the first factor is added to the number of factors represented by the exponent of the second factor:

Three factors
plus two factors
equal five factors

$$4^3 \cdot 4^2 = 4^5$$

This addition of exponents may be used when several powers are multiplied together:

$$x^2 \cdot x^3 \cdot x^2 = x^7$$

Remember, however, that **exponents may be added only for powers with the same base.**

MULTIPLYING WITH ZERO OR NEGATIVE EXPONENTS

The exponent addition rule is valid even when the exponent is zero or negative. For example:

$$2^3 \cdot 2^0 = 2^3$$

and $3^3 \cdot 3^{-2} = 3^1$

WHEN MULTIPLYING POWERS WITH THE SAME BASE—ADD EXPONENTS

Using the expanded factor form, expand each factor, then determine the exponent of the answer:

$$x^3 \cdot x^4 = \underbrace{x \cdot x \cdot x}_{x^3} \cdot \underbrace{x \cdot x \cdot x \cdot x}_{x^4} = x^7$$

Using the shortcut, simply add exponents:

$$x^3 \cdot x^4 = x^{3+4} = \boxed{x^7}$$

Example 1: (Multiplying and Dividing Powers)
❏ Simplify the following terms using the rule for multiplication of powers.

$$3^2 \cdot 3^5 = 3^{2+5}$$

$$= \boxed{3^7}$$

$$x^3 \cdot x^3 = x^{3+3}$$

$$= \boxed{x^6}$$

Example 2: (Multiplying and Dividing Powers)
❏ Simplify each multiplication problem below, using the rule for multiplication of powers.

$$12 \cdot 12^3 = 12^{1+3}$$

$$= \boxed{12^4}$$

$$5^3 \cdot 5^{-2} = 5^{3-2}$$

$$= 5^1 \text{ or } \boxed{5}$$

$$2^2 \cdot 3^2 \cdot 2^3 = 2^{2+3} \cdot 3^2$$

$$= \boxed{2^5 \, 3^2}$$

---

* This is consistent with the fact that any factor to the zero power is equal to one. In this example $2^3 \cdot 2^0 = 2^3$, just as $2^3 \cdot 1 = 2^3$.

WHEN DIVIDING POWERS WITH THE SAME BASE—SUBTRACT EXPONENTS

Using the long method, expand each factor, cancel terms, and determine the exponent of the answer:

$$\frac{x^5}{x^2} = \frac{x \cdot x \cdot x \cdot \cancel{x} \cdot \cancel{x}}{\cancel{x} \cdot \cancel{x}} = \boxed{x^3}$$

Using the shortcut, simply subtract exponents:
(Always subtract the denominator exponent from the numerator exponent)

$$x^{5-2} = \boxed{x^3}$$

DIVIDING POWERS

Factors in the numerator may be divided out by the same factors in the denominator, as shown in the box to the left.

Instead of expanding the factors and dividing out similar factors, you may simply subtract the exponent of the denominator from the exponent of the numerator. This produces the same result and is a much quicker process.

DIVIDING WHEN THERE ARE NEGATIVE EXPONENTS IN THE DENOMINATOR

If there is a negative exponent in the denominator, you may wish to move it to numerator and change its sign to positive,* and proceed as usual. This eliminates what is a confusing step to many people—subtracting a negative number. For example, suppose the problem is $2^2 \div 2^{-3}$; either of the following methods may be used to simplify the problem:

$$\frac{2^2}{2^{-3}} = 2^2 \cdot 2^3 = \boxed{2^5}$$

$$\frac{2^2}{2^{-3}} = 2^{2-(-3)} = 2^{2+3}$$
$$= \boxed{2^5}$$

Example 3: (Multiplying and Dividing Powers)
❏ Simplify the following terms using the rule for division of powers.

$$\frac{a^3}{a^2} = a^{3-2}$$
$$= a^1 \text{ or } \boxed{a}$$
$$\frac{9^5}{9^3} = 9^{5-3}$$
$$= \boxed{9^2}$$

Example 4: (Multiplying and Dividing Powers)
❏ Calculate the answer to each of the following problems using the rules for multiplication or division of exponents.

$$\frac{2^3 \cdot 3^4}{2^2} = 2^{3-2} \cdot 3^4$$
$$= 2^1 \cdot 3^4$$
$$= \boxed{162}$$

$$\frac{5^3 \cdot 7^2 \cdot 5^2}{5} = 5^{3+2-1} \cdot 7^2$$
$$= 5^4 \cdot 7^2$$
$$= \boxed{30,625}$$

---

* Refer to Section 13.2 for a review of changing the sign of the exponent.

PRACTICE PROBLEMS 13.4: Multiplying and Dividing Powers

❏ Simplify each of the following terms, using the rules for multiplication and division of powers. (No fractions.)

1. $\dfrac{17x^3}{x}$ = _____

2. $(4^2)(5^{-2})(5)$ = _____

3. $\dfrac{x^2 \cdot x^0}{3 \cdot x^3}$ = _____

4. $3x^2 \cdot 3x^{-5}$ = _____

5. $\dfrac{18x^{-3}\,y^{-5}z^3}{9x\,yz^2}$ = _____

6. $\dfrac{2x^{-3}\,y\,z^2}{x\,y^3}$ = _____

❏ Complete the following problems using the rules for multiplication and division of powers.

7. $\dfrac{6^3 \cdot 7}{6^2}$ = _____

8. $\dfrac{(8)(3^2)(8)}{3^3}$ = _____

9. $\dfrac{4^{-2} \cdot 2^6}{5^{-2}}$ = _____

10. $\dfrac{2^0 \cdot 2^7 \cdot 5}{5^3 \cdot 3}$ = _____

13.5 POWERS OF TEN

SUMMARY

1. When **multiplying by a power of 10,** move the decimal point to the **right** the same number of places as the power of ten (or the number of zeros in the expanded power of ten).

$$(3.6)(10^2) = 360.$$

power of
ten is 2

$$1\ 2$$

move the decimal
2 places to the right

2. When **dividing by a power of 10,** move the decimal point to the **left** the same number of places as the power of ten.

$$\frac{184.4}{10^2} = 184.2$$

$$2\ 1$$

power of
ten is 2

move the decimal
2 places to the left

MULTIPLYING BY A POWER OF TEN

When multiplying a number by ten or a power of ten, the result is simply **a change in the decimal point position to the right** of that in the original number. For example:

$$14 \times 10 = 140.$$

$$\text{or } 26.5 \times 10^2 = 2650.$$

In the first instance, the decimal point moved one place to the right, and in the second case, the decimal point moved two places to the right. In these two examples, notice that the decimal point move corresponded with the power of ten—power of one, decimal move of one; power of two, decimal move of two. In fact, this is always the case when multiplying by a power of ten. **The exponent (1, 2, 3, etc.) indicates how many places the decimal point should be moved to the right.**

If the power of ten is written in expanded form, such as 10, 100, 1000, etc., **the number of zeros indicates how many places to the right the decimal point must be moved.** For example:

$$0.56 \times 100 = 0.56.$$

$$= 56.0$$

(*Hint*: As an aid in deciding whether to move the decimal point to the right or left when multiplying, remember that multiplication of a whole number **increases** that number. A decimal point move to the right also **increases** the number.)

Example 1: (Powers of Ten)
❑ Use only decimal point moves to complete the following multiplication problems.

$$(1.42)(10^2) = 142.$$

$$= \boxed{142}$$

$$(0.862)(10^3) = 0862.$$

$$= \boxed{862}$$

Example 2: (Powers of Ten)
❑ Complete the multiplication problems shown below using decimal point moves.

$$(1.05)(10,000) = 10500.$$

$$= \boxed{10,500}$$

$$(2.95)(1,000,000) = 2950000.$$

$$= \boxed{2,950,000}$$

$$(0.00032)(100) = 000.032$$

$$= \boxed{0.032}$$

Example 3: (Powers of Ten)
❑ The division problems given below are to be completed using decimal point moves only.

$$\frac{362}{10^2} = 3.62$$

$$= \boxed{3.62}$$

$$\frac{1,876,000}{10^6} = 1.876000$$

$$= \boxed{1.876}$$

Example 4: (Powers of Ten)
❑ Use decimal point moves to complete the division problems shown below.

$$\frac{92,640}{1000} = 92.640$$

$$= \boxed{92.64}$$

$$\frac{325}{10,000} = .0325$$

$$= \boxed{0.0325}$$

DIVIDING BY A POWER OF TEN*

Any number divided by a power of ten results in **a decimal point move to the left of that in the original number**. For example:

$$17.62 \div 10^3 = 0.01762$$

The exponent of the power of ten indicates how many places to move the decimal point to the left.

If the power is written in expanded form (such as 10, 100, 1000, etc.), the number of zeros indicates how many places to the left the decimal point must be moved. For example:

$$867.4 \div 100 = 8.674$$

(*Hint*: To remember which direction to move the decimal point, remember that division by a whole number results in a **smaller** number. A decimal point move to the left also results in a **smaller** number.)

---

* Multiplying by a negative power of ten is the same as dividing by the same positive power. For example, 3×10^{-2} is equivalent to $3 \div 10^2$. For more practice multiplying with negative exponents, refer to Section 13.2.

PRACTICE PROBLEMS 13.5: Powers of Ten

❏ Complete the following multiplication and division problems using decimal moves only.

1. 16.3×10^3 = _____

6. $\dfrac{24.09}{100}$ = _____

2. $(0.0061)(10^2)$ = _____

7. $(0.99)(10^4)$ = _____

3. $8220 \div 10$ = _____

8. $(3,600)(100)$ = _____

4. $(4.62)(10,000)$ = _____

9. $4,260,000 \div 10^6$ = _____

5. $62 \div 1000$ = _____

10. $\dfrac{787}{10,000}$ = _____

13.6 SCIENTIFIC NOTATION

> ### SUMMARY
>
> 1. **Scientific notation** is a method of writing numbers such that there is a number (always between 1 and) times a power of ten.
>
> 2. **To put a number in scientific notation:**
>
>   • Place a decimal point after the first nonzero digit. This is called the "standard position" for the decimal point.
>
>   • Count the number of places from the standard position to the original decimal point. This represents the exponent of the power of ten.
>
>   • If the move from the standard position to the original decimal point was to the right, the exponent is positive; if the move was to the left, the exponent is negative.
>
> $$3,420,000. = 3.42 \times 10^6$$
>
> |←— 6 places —→|
>
> *standard position of decimal point*   *original position of decimal point*   *positive exponent*
>
> 3. When **multiplying or dividing numbers in scientific notation**:
>
>   • Separate the factors of ten and simplify the expression.
>
>   • Multiply or divide the other factors, as indicated
>
>   • Leave the answer in scientific notation or take it out of scientific notation, as desired.
>
> 4. **To take a number out of scientific notation,** simply multiply by the power of ten as indicated. A positive exponent indicates a decimal move to the right. A negative exponent indicates a decimal move to the left.

Numbers used in various fields of science are often very large or very small. For this reason, a method of writing numbers was developed to make working with such numbers less cumbersome. This method is called **scientific notation.**

Using scientific notation, the decimal point is rearranged so that it is always after the first nonzero digit. This position of the decimal point is called the **standard position.** The number, now in standard for, is multiplied by a positive or negative power of ten to compensate for moving the decimal place.

Beginning at the standard position, a move to the right toward the original decimal point indicates a positive exponent of ten; a move to the left indicates a negative exponent of ten. The number of the exponent (1, 2, 3, etc.) depends on how many places are counted from the standard position to the original position of the decimal point. The examples given on this page illustrate how to write a number in scientific notation.

PUTTING A NUMBER INTO SCIENTIFIC NOTATION

$$2.675. = 2.675 \times 10^3$$

① First, place the decimal point in the "standard position"—after the first nonzero number.

② Next, count the number of places from the standard position to the original decimal point. A move to the right indicates a positive exponent of ten, and a move to the left indicates a negative exponent of ten.

③ Write the number in scientific notation. In this example, the exponent of ten is a positive three, since the decimal move from the standard position to the original position was three places to the right.

Example 1: (Scientific Notation)

❏ Write the following numbers in scientific notation: 14,500; 0.02; 2,970,000; 0.0035.

First, find the decimal point in standard position, then determine the number and sign of the power of ten exponent:

$$1.4500 = \boxed{1.45 \times 10^4}$$

4 places
to the right

$$0.02 = \boxed{2.0 \times 10^{-2}}$$

2 places
to the left

$$2.970000 = \boxed{2.97 \times 10^6}$$

6 places
to the right

$$0.003.5 = \boxed{3.5 \times 10^{-3}}$$

3 places
to the left

Example 2: (Scientific Notation)
❏ Take the following numbers out of scientific notation:

3.614×10^5; 9.06×10^{-3}; 8.426×10^6

$$3.614 \times 10^5 = 3.61400 = \boxed{361,400}$$

*A positive exponent of 5
indicates a decimal point
move of 5 places to the right*

$$9.06 \times 10^{-3} = 009.06 = \boxed{0.00906}$$

*A negative exponent of 3
indicates a decimal point
move of 3 places to the left*

$$8.426 \times 10^6 = 8.426000 = \boxed{8,426,000}$$

*A positive exponent of 6
indicates a decimal point
move of 6 places to the right*

TAKING A NUMBER OUT OF SCIENTIFIC NOTATION

Should you desire to take a number out of scientific notation and express it as a regular decimal number, the process is simple. Merely multiply by the accompanying power of ten. If the power of ten has a **positive exponent, move the decimal point to the right** the number of places indicated by the exponent. If the power of ten has a **negative exponent, move the decimal point to the left** the number of places indicated by the exponent.

Example 3: (Scientific Notation)
❏ Complete the following calculation, keeping factors in scientific notation until the final answer.

$$\frac{(2.3 \times 10^6)(8.24 \times 10^{-3})}{1.82 \times 10^2}$$

First, separate and simplify the powers of 10 :

$$\frac{(2.3)(8.24)}{1.82} \times 10^{6-3-2}$$

or $\dfrac{(2.3)(8.24)}{1.82} \times 10^1$

Then complete the calculation of the remaining factors:

$$10.41 \times 10^1 = \boxed{104.1}$$

CALCULATING WITH SCIENTIFIC NOTATION

Sometimes the numbers to be calculated are too large for use with a handheld calculator (which only displays 8 to 10 digits). When this occurs, you can put one or more numbers in scientific notation so the calculation can be completed. It is not necessary that all the numbers be placed in scientific notation.

To make calculations with numbers given in scientific notation, first separate the powers of ten from the other factors and simplify, using the rules of multiplication and division of powers*. Then multiply or divide the other factors. Example 3 illustrates how this is done.

---

* For a discussion of these rules and examples, refer to Section 13.4.

PRACTICE PROBLEMS 13.6: Scientific Notation

❑ Write the following numbers in scientific notation.

1. 16,210 = _____

3. 0.0028 = _____

2. 145 = _____

4. 4,640,000 = _____

❑ Take the numbers shown below out of scientific notation.

5. 2.15×10^3 = _____

7. 8.915×10^6 = _____

6. 6.77×10^{-5} = _____

❑ Complete the following calculations, keeping factors in scientific notation until the final answer.

8. $(7.62 \times 10^2)(3.04 \times 10^3)$ =

10. $(2.6 \times 10^{-2})(1.47 \times 10^6)$ =

ANS _____

ANS _____

9. $\dfrac{(5.32 \times 10^6)(1.18 \times 10^3)}{4.69 \times 10^3}$

ANS _____

14 *Rounding and Estimating*

Complete and score the following skills test. Each section should be scored separately in the box provided to the right. A score of 4 or above indicates that you are sufficiently strong in that concept. A score of 3 or below indicates a review of that section is advisable.

14.1 Rounding

Number Correct

❑ Round the following numbers to the nearest tenth.

1. 19.845 = _____

2. 0.187 = _____

❑ Round to the nearest hundred.

3. 46,230 = _____

4. 82.056 = _____

❑ Round to the nearest ten thousand.

5. 640,960 = _____

14.2 Significant Figures

Number Correct

❑ Round the following numbers to two significant figures.

1. 613,400 = _____

3. 0.177 = _____

2. 89,125 = _____

4. 0.1052 = _____

14.2 Significant Figures—Cont'd

5. 3,194,000 = _____

14.3 Estimating

Number
Correct

❑ Estimate the answers to the problems shown below. (Do not use a calculator as part of the estimation.)

1. (12)(12)(28) = _____

4. $\dfrac{(3)(8600)}{(8)(712)}$ = _____

2. (0.785)(62)(62) = _____

5. (18)(25)(38) = _____

3. $\dfrac{(42)(1500)}{95}$ = _____
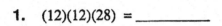

14.1 ROUNDING

1. In order to round a number you must know the **place value system**.

   To the left of the decimal point are the following place values:

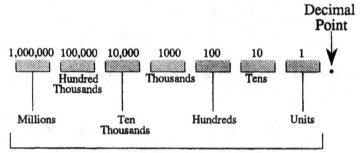

Values Greater Than One

   To the right of the decimal point are the following place values:

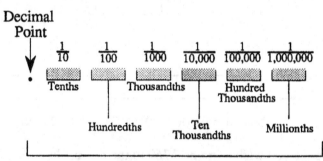

Values Less Than One

2. Rounding procedures depend on whether the **number to the right** of the "rounding place" is less than 5, equal to 5, or greater than 5. The procedures also vary depending on whether the entire number being rounded is less than one or greater than one.

Rounding may be necessary when estimating an answer or when giving an answer to a specified number of significant figures.

In order to round, it is essential to know the name and value of each place in the place value system. A diagram of the most common place values is given in the box to the right. Note that place values less than one include a "th" on the end of the word. This helps distinguish these place values from those greater than one.

WHEN ROUNDING NUMBERS TO THE LEFT OF THE DECIMAL POINT

The rounding procedures are slightly different for numbers rounded to the left of the decimal point than for numbers rounded to the right of the decimal point. **When rounding to the left of the decimal point:**

1. Find the "rounding place" as stated by the problem. The number in that position will either stay the same or will be increased by one, **depending on the size of the number to the right of the "rounding place."**

2. **If the number to the right is 5 or greater,** increase the rounding position by one and replace all numbers to the right with zeros. For example,

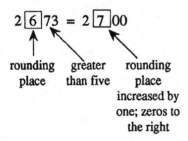

$$2\boxed{6}73 = 2\boxed{7}00$$

rounding place   greater than five   rounding place increased by one; zeros to the right

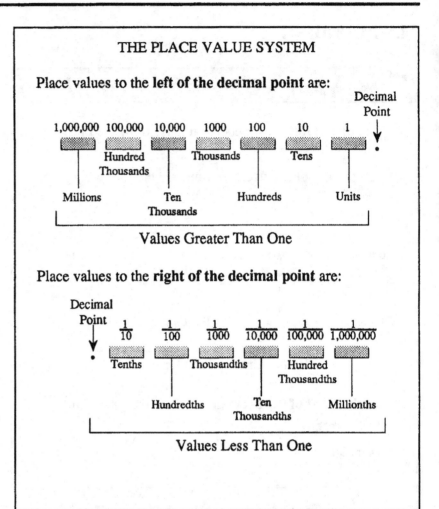

THE PLACE VALUE SYSTEM

Place values to the **left of the decimal point** are:

Decimal Point

1,000,000   100,000   10,000   1000   100   10   1

Hundred Thousands   Thousands   Tens

Millions   Ten Thousands   Hundreds   Units

Values Greater Than One

Place values to the **right of the decimal point** are:

Decimal Point

$\frac{1}{10}$ $\frac{1}{100}$ $\frac{1}{1000}$ $\frac{1}{10,000}$ $\frac{1}{100,000}$ $\frac{1}{1,000,000}$

Tenths   Thousandths   Hundred Thousandths

Hundredths   Ten Thousandths   Millionths

Values Less Than One

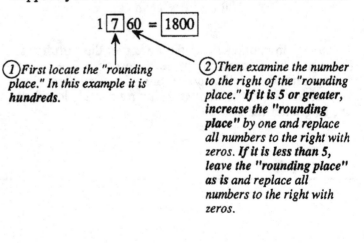

ROUNDING NUMBERS GREATER THAN ONE

Suppose you wish to round 1760 to the nearest hundred:

$$1\boxed{7}60 = \boxed{1800}$$

①*First locate the "rounding place." In this example it is hundreds.*

②*Then examine the number to the right of the "rounding place." If it is 5 or greater, increase the "rounding place" by one and replace all numbers to the right with zeros. If it is less than 5, leave the "rounding place" as is and replace all numbers to the right with zeros.*

Example 1: (Rounding)
❏ Round the following numbers to the nearest thousand:

$$4\boxed{8}, 620 = \boxed{49{,}000}$$

$$\boxed{1}\,150 = \boxed{1000}$$

$$7{,}22\boxed{4}, 860 = \boxed{7{,}225{,}000}$$

If the number to the right is less than 5, leave the "rounding place" as is and replace all numbers to the right with zeros. For example,

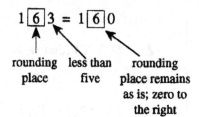

ROUNDING NUMBERS LESS THAN ONE

Suppose you wish to round 0.639 to the nearest tenth:

$$0.\boxed{6}\,39 = \boxed{0.6}$$

Locate "rounding place"   *Round according to the same "5 or greater" rule,* **then drop all numbers to the right.**

WHEN ROUNDING NUMBERS TO THE RIGHT OF THE DECIMAL POINT

There is only one difference in rounding procedures for place values to the right of the decimal point. Instead of replacing all numbers to the right with zeros, simply **drop the numbers to the right of the "rounding place".** Example 2 illustrates this procedure.

Example 2: (Rounding)
❏ Round the following numbers to the nearest tenth:
20.54; 419.654; 0.0825

$$20.\boxed{5}\,4 = \boxed{20.5}$$

$$419.\boxed{6}\,54 = \boxed{419.7}$$

$$0.\boxed{0}\,825 = \boxed{0.1}$$

PRACTICE PROBLEMS 14.1: Rounding

❏ Round the following numbers to the place value indicated.

1. 3762 (tens) = _____

2. 697,423 (thousands) = _____

3. 3,589,265 (millions) = _____

4. 192.172 (hundredths) = _____

5. 44,713 (ten thousands) = _____

6. 0.2556 (thousandths) = _____

7. 29.76 (units) = _____

8. 186.37 (tens) = _____

9. 4,680,990 (hundred thousands) = _____

10. 3.17456 (ten thousandths) = _____

14.2 SIGNIFICANT FIGURES

SUMMARY

1. **Significant figures** are a means of reflecting the accuracy of a measurement.

2. **When determining the number of significant figures** in a number:

    - The digits 1 through 9 are always signifcant figures.

    - Zeros are sometimes significant figures and sometimes not, depending on where they are located in the number.

    - The digits shown in scientific notation are always significant figures.

3. The answer to a calculation should include no more significant figures than the **least number of significant figures** in the numbers used for that calculation.

    For example, if the numbers 1.9, 217, and 1526 are used in a calculation, the answer should contain no more than two signficant figures (1.9 has the least number of significant figures—two).

Every measurement is approximate. That is, it is only accurate to a certain limit and depends on the instrument used, the conditions during measurement, the accuracy of the reading, and other such factors.

The concept of **significant figures** was developed to indicate the relative accuracy of a particular measurement.

To determine how many significant figures a number has, **the following guidelines should be considered:**

1. Digits 1 through 9 are always considered significant figures.*

2. Zeros are sometimes considered significant figures, depending on where they are located in the number. Refer to the box to the right.

3. The digits given in scientific notation are always considered significant figures.

ZEROS—ARE THEY SIGNIFICANT FIGURES OR NOT?

Zeros always present the greatest difficulty when determining significant figures.

- **Zeros between other digits (1-9) are always significant figures:**

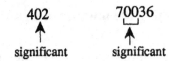

- **Zeros used to locate the decimal point are not significant figures:**

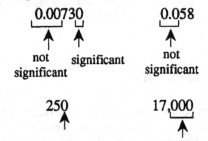

If these zeros are used to locate the decimal point, they are not significant figures. If they are accurate measurement readings, they are significant figures. Written in this form, there is no way for you to discern which is which. Writing a number in scientific notation eliminates this confusion.

- **The zeros included as scientific notation digits are always significant figures:**

$$2.50 \times 10^2 \qquad 1.700 \times 10^4$$

One zero is significant. A total of 3 significant figures.

Two zeros are significant. A total of 4 significant figures.

---

* This refers to measurements, not answers to a calculation. Answers to a calculation are rounded to a certain number of significant figures, depending on the numbers used in the calculation.

Example 1: (Significant Figures)
❑ Determine the number of significant figures in each of the following numbers: 0.708; 0.004030; 7002; 62.005; and 72.0.

0.<u>708</u>— Three significant figures.

0.004<u>030</u>— Four significant figures. (The zero to the far right is considered significant. It does not locate the decimal point and is considered a reading accurate to that place value.)

<u>7002</u>— Four significant figures.

<u>62.005</u>— Five significant figures.

<u>72.0</u>— Three significant figures. (Again, the zero to the right is considered an accurate reading. It is not necessary to locate the decimal point.)

ROUNDING TO A GIVEN NUMBER OF SIGNIFICANT FIGURES

The process of rounding to a specified number of significant figures is not much different than rounding as described in Section 14.1. However, rather than rounding to a given place value you must first locate the desired significant figure as the "rounding point". Then proceed according to usual rounding rules.

Example 2: (Significant Figures)
❑ Round each of the following numbers to three significant figures: 1.6834; 0.040932; 423,792; 31.7215.

Find the three significant figures and round off all other digits.

$$1.6834 = 1.68$$

$$0.040932 = 0.0409$$

$$423,792 = 424,000$$

$$31.7215 = 31.7$$

CALCULATIONS AND SIGNIFICANT FIGURES

When several measurements are used in a calculation, the resulting answer can be no more accurate than the individual numbers used in that calculation. This is similar to the maxim, "a chain is only as strong as its weakest link."

Once you have determined the answer to a calculation, examine the numbers used in the calculation to **find the number with the least significant figures**. Then round your answer to that many significant figures.

Example 3: (Significant Figures)
❑ Complete the following problem and round the answer to the correct number of significant figures.

$$(16.75)(1.5)(252) = 6331.5$$

The factor with the least number of significant figures is 1.5—two significant figures. Therefore the answer should be rounded to two significant figures:

$$6331.5 = \boxed{6300}$$

PRACTICE PROBLEMS 14.2: Significant Figures

❏ Determine the number of significant figures in each of the following numbers.

1.   793,324 = _____

3.   0.005617 = _____

2.   4.306 = _____

4. 7.00×10^6 = _____

❏ Round each of the following numbers to three significant figures.

5.   72.0032 = _____

7.   0.02709 = _____

6.   422.45 = _____

8.   49,823 = _____

❏ Complete the calculations given below. Round the answer to the correct number of significant figures.

9. $\dfrac{(9.72)(2725)}{0.55}$ = _____

10.   (6808)(34.12)(0.0015) = _____

14.3 ESTIMATING

SUMMARY

For a rough estimate of a multiplication calculation:

1. Round all numbers to one significant figure.

2. Multiply significant figures.

3. Add the total number of zeros in the factors to the product from Step 2.

4. Adjust the decimal point in the answer for any decimal fractions of the factors.

For a rough estimate of a calculation that includes both multiplication and division:

1. Round all numbers to one significant figure.

2. Adjust the decimal point of any factors less than one and make a corresponding decimal point move in the numerator or denominator of the fraction, depending on the location of the factor less than one.

3. Divide out as many zeros as possible.

4. Divide out factors, where possible.

5. Round the numerator and denominator as needed to make an easy division problem, then divide.

ROUGH ESTIMATES

Many water and wastewater calculations involve strictly multiplication and division. Whether the calculation is to be completed by hand or with a calculator, it is wise to have a rough idea of the answer. This will help prevent obvious decimal errors. The steps described in detail below are intended to be done quickly with a paper and pencil. Some of them can be done mentally. Estimating skills improve greatly with practice.

ESTIMATING PRODUCTS

For a quick estimate of the problems involved:

1. **Round all numbers** to one significant figure*.

2. **Multiply significant figures.** (If several figures are to be multiplied, it may be necessary to round during this process.)

3. **Add the total number of zeros** (from all numbers greater than one) to the end of the product obtained in Step 2.

4. For any factor less than one, **adjust the decimal point in the answer** the same number of places to the left as the decimal point in that factor.

Examples 1 and 2 illustrate this estimation process.

Example 1: (Estimating)
❑ Estimate the answer to the following problem.

$$(67.5)(7.48)$$

First, round each term to one significant figure, then multiply single digits and add zeros.

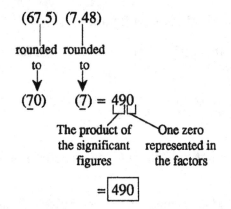

The product of the significant figures

One zero represented in the factors

$$= \boxed{490}$$

Example 2: (Estimating)
❑ Estimate the answer to the following problem.

$$(2.78)(60)(1440)(7.48)$$

Round each term, then multiply single digits, then add zeros:

$$(2.78)(60)(1440)(7.48)$$

$$(3) \quad (60) \quad (1000) \quad (7)$$

$$180 \times 1000 \times 7$$

rounded
to $\quad 200 \times 1000 \times 7 = 14\,00000$

4 zeros from rounded factors

$$= \boxed{1,400,000}$$

---

* Refer to Section 14.2 for a discussion of significant figures.
** For a discussion of decimal fractions, refer to Chapter 4.

Example 3: (Estimating)
❑ Estimate the answer to the following problem.

$$\frac{790,500}{43,560}$$

First, round both terms, then divide out zeros where possible:

$$\frac{790,500}{43,560} \quad \text{Rounded to} \quad \longrightarrow \quad \frac{800,000}{40,000}$$

$$= \frac{80}{4}$$

No further rounding is necessary to determine the estimated answer:

$$= \boxed{20}$$

Example 4: (Estimating)
❑ Estimate the answer to the problem given below:

$$\frac{(0.78)(65)}{210}$$

First round all factors to one significant figure:

$$\frac{(0.8)(70)}{200}$$

Adjust the decimal point in the decimal fraction (one move to the right) and make a corresponding move in the denominator:

$$\frac{(0.8)(70)}{200.0}$$

Now, divide out zeros and factors, if possible:

$$\frac{\overset{4}{(8)}(70)}{\underset{1}{2000}} = \frac{28}{100} = \boxed{0.28}$$

WHEN CALCULATIONS INCLUDE DIVISION

In the examples thus far, the calculations have included only multiplication. When division is part of the calculation:

1. **Round all numbers** to one significant figure.

2. **For any factor less than one, move the decimal point** to the right unit the significant figure is in the units place. If this factor is in the numerator, make a corresponding decimal point change in any one factor in the denominator. If the factor is in the denominator, make a corresponding decimal point change in the numerator.

3. **Divide out as many zeros** as possible.

4. **Divide factors** where possible.

5. **Round numerator and denominator** as needed to make an easy division problem, then divide.

PRACTICE PROBLEMS 14.3: Estimating

❏ Estimate the answers to the following multiplication problems.

1.　(0.785)(40)(40) = _____

4.　(3.05)(60)(1440)(7.48) = _____

2.　(14,850)(86.2)(47) = _____

5.　(0.785)(60)(60)(25)(7.48) = _____

3.　(2800)(0.01)(8.34) = _____

6.　(40)(15)(8)(7.48) = _____

❏ Estimate the answers to the multiplication and division problems given below.

7.　$\dfrac{(4)(2.5)}{120}$ = _____

9.　$\dfrac{2,100,000}{(0.785)(80)(80)}$ = _____

8.　$\dfrac{(22)(16,680)}{33,000}$ = _____

10.　$\dfrac{(12)(60)(1440)(0.22)}{(3)(2.5)(43)(7.48)}$ = _____

15 *Dimensional Analysis*

Complete and score the following skills test. Each section should be scored separately in the box provided to the right. A score of 4 or above indicates that you are sufficiently strong in that concept. A score of 3 or below indicates a review of that section is advisable.

15.1 The Basics

Number Correct

❑ Use dimensional analysis to determine the units of the answer for each problem given below.

1. (3 ft)(2.5 ft)(2 ft/sec) =

ANS _____

2. (0.785)(90 ft)(90 ft) =

ANS _____

3. (0.8 cu ft/sec)(60 sec/min) =

ANS _____

4. (2.25 cu ft/sec)(60 sec/min)(7.48 gal/cu ft) =

ANS _____

5. $\dfrac{(10 \text{ ft})(8 \text{ ft})(x \text{ ft})}{3 \text{ min}}$ =

ANS _____

15.2 Problems Involving Complex Fractions

❑ Use dimensional analysis to determine the units of the answer for each problem given below.

1. $\dfrac{4500 \text{ gpm}}{60 \text{ sec/min}}$ =

ANS _____

2. $\dfrac{15 \text{ gpm/sq ft}}{7.48 \text{ gal/cu ft}}$ =

ANS _____

3. $\dfrac{400 \text{ cu ft}}{3.4 \text{ cu ft/day}}$ =

ANS _____

4. $\dfrac{130,900 \text{ gal}}{115,833 \text{ gph}}$ =

ANS _____

5. $\dfrac{853.5 \text{ cu ft}}{27 \text{ cu ft/cu yds}}$ =

ANS _____

15.1 DIMENSIONAL ANALYSIS—THE BASICS

SUMMARY

When using dimensional analysis to check the set-up of a mathematical calculation, three basic concepts are involved:

1. Units written in abbreviated or horizontal form should be rewritten in a **vertical format**. For example:

$$\text{cfs} \longrightarrow \frac{\text{cu ft}}{\text{sec}}$$

$$\text{or} \quad \text{gal/cu ft} \longrightarrow \frac{\text{gal}}{\text{cu ft}}$$

2. Any unit which is a **common factor to both the numerator and denominator** of a fraction may be divided out. For example:

$$20 \, \frac{\text{cu ft}}{\cancel{\text{sec}}} \quad \text{x} \quad 60 \, \frac{\cancel{\text{sec}}}{\text{min}}$$

3. An **exponent of a unit** indicates how may times that unit is to be multiplied together. For example:

$$\text{ft}^3 = (\text{ft})(\text{ft})(\text{ft})$$

Sometimes it is necessary to write terms with exponents in expanded form, while other times it is advantageous to keep the unit in exponent form. This choice depends on which other units are part of the calculation and how these units might divide out.

WHEN TO USE DIMENSIONAL ANALYSIS

Dimensional analysis is a very valuable tool used to check the accuracy of the mathematical set-up for a particular problem. It is recommended that dimensional analysis be used to verify a mathematical set-up rather than to construct one.

All examples and practice problems presented in this chapter have been adapted from calculations found throughout the applied math texts.

DIVIDE OUT SIMILAR UNITS

In dimensional analysis units in the numerator are divided out with similar units in the denominator, such as:

$$4 \,\cancel{ft} \times \frac{12 \text{ in.}}{\cancel{ft}}$$

These terms may be divided out just as factors are divided out when making calculations.

In order to complete the division of units, it is important that all units be written in the same format:

horizontal or vertical

| gal/cu ft | $\dfrac{\text{gal}}{\text{cu/ft}}$ |

The vertical format is used almost exclusively in dimensional analysis.

In addition, any abbreviations should be written in the vertical format, such as:

$$\text{gpm} \longrightarrow \frac{\text{gal}}{\text{min}}$$

$$\text{or} \quad \text{psi} \longrightarrow \frac{\text{lbs}}{\text{sq in.}}$$

Example 1: (Dimensional Analysis—Basics)
❑ Check the mathematical set-up given below, using dimensional analysis. Do the desired terms of the answer match with the math set-up as shown?

$$(0.785)(40 \text{ ft})(40 \text{ ft}) = \text{ft}^2$$

Analyze only the units of the problem:

$$(\text{ft})(\text{ft}) = \text{ft}^2$$

$$\boxed{\text{ft}^2} = \boxed{\text{ft}^2}$$

*The units on the left side of the equation **match** the units on the right side of the equation.*

Example 2: (Dimensional Analysis—Basics)
❑ The mathematical set-up for a problem is given below. Do the units of the problem result in the desired units of the answer?

$$(1.2 \text{ cfs})(60 \text{ sec/min})(7.48 \text{ gal/cu ft}) = \text{gpm}$$

First express all units in a vertical format, then evaluate units:

$$\frac{\cancel{\text{cu ft}}}{\cancel{\text{sec}}} \cdot \frac{\cancel{\text{sec}}}{\text{min}} \cdot \frac{\text{gal}}{\cancel{\text{cu ft}}} = \frac{\text{gal}}{\text{min}}$$

$$\boxed{\frac{\text{gal}}{\text{min}}} = \boxed{\frac{\text{gal}}{\text{min}}}$$

Units on the left match
the units on the right

Example 3: (Dimensional Analysis—Basics)
❏ Verify the following problem set-up using dimensional analysis. Is the set-up correct according to the desired units of the answer?

$$\frac{(127,170 \text{ ft}^3)(20 \text{ ft})}{0.785} = \text{ft}^2$$

Review the units of the problem:

$$(\text{ft}^3)(\text{ft}) = \text{ft}^2$$

$$\lfloor \text{ft}^4 \rfloor \neq \lfloor \text{ft}^2 \rfloor$$

*The units on the left side of the equation **do not match** the desired units. There is an error in the mathematical set-up.*

After reconsidering the math set-up, the error was detected. The set-up should have been:

$$\frac{(127,170 \text{ ft}^3)}{(0.785)(20 \text{ ft})} = \text{ft}^2$$

Example 4: (Dimensional Analysis—Basics)
❏ Using dimensional analysis to check the mathematical set-up shown below, do the units of the problem result in the desired units of the answer?

$$\frac{(0.785)(50 \text{ ft})(50 \text{ ft})(3 \text{ ft})}{2} = \text{ft}^2$$

Examine the units of the problem:

$$(\text{ft})(\text{ft})(\text{ft}) = \text{ft}^2$$

$$\text{ft}^3 \neq \text{ft}^2$$

The units on the left and right do not match. Therefore there is something wrong with the math set-up. In fact, there is an extra "ft" factor on the left side that does not belong there.

WHEN UNITS INCLUDE EXPONENTS

Units written with exponents,* such as ft^2, can be left as is or put in expanded form, (ft)(ft), depending on other units in the calculation.

In any case, be sure that square and cubic terms are expressed uniformly —using either English abbreviations (sq ft, cu ft) or international abbreviations (ft^2, ft^3). The latter system is preferred in dimensional analysis. Example 3 illustrates this type of calculation.

DIMENSIONAL ANALYSIS FOR mg/*L* TO lbs/day CALCULATIONS

One very common calculation used in water and wastewater treatment calculations is mg/*L* to lbs/day calculations.** Using the units normally associated with the calculation, dimensional analysis does not appear to work:

$$\frac{\text{mg}}{L} \cdot \frac{\text{MG}}{\text{day}} \cdot \frac{8.34 \text{ lbs}}{\text{gal}} = \frac{\text{lbs}}{\text{day}}$$

However, remember that mg/*L* and parts per million (ppm) are considered interchangeable. "Parts per million parts" can apply to any units, and lbs per million lbs are preferred for chemical concentration problems (rather than gal per million gallons). The dimensional analysis is therefore:

$$\frac{\text{lbs}}{\cancel{\text{mil lbs}}} \cdot \frac{\cancel{\text{MG}}}{\text{day}} \cdot \frac{8.34 \, \cancel{\text{lbs}}}{\cancel{\text{gal}}} = \frac{\text{lbs}}{\text{day}}$$

The equation is written as (mg/*L*)(MGD)(8.34 lbs/gal) = lbs/day, because it corresponds better with plant data.

---

* For a discussion of exponents, refer to Chapter 13.
** Refer to Chapter 3 of the applied mathematics texts for a discussion of mg/*L* to lbs/day calculations.

PRACTICE PROBLEMS 15.1: Dimensional Analysis—The Basics

❑ Use dimensional analysis to determine the units of the answers in the problems below.

1.   (0.785)(ft)(ft)(ft) =

ANS _____

2.   (120 cu ft/min)(1440 min/day) =

ANS _____

3. $\dfrac{(8 \text{ ft})(10 \text{ ft})(x \text{ ft})}{300 \text{ sec}}$ =

ANS _____

❑ Verify the mathematical set-up for each problem below using dimensional analysis. Do the desired terms of the answer match with the set-up as shown? What should the set-up be in order to achieve the terms of the answer?

4.   (1.6 fpm)(60 $\dfrac{\text{sec}}{\text{min}}$) = fps

ANS _____

5.   (70 in.) $\dfrac{(1 \text{ ft})}{12 \text{ in}}$ $\dfrac{(0.3048 \text{ m})}{\text{ft}}$ = m

ANS _____

15.2 DIMENSIONAL ANALYSIS AND COMPLEX FRACTIONS

SUMMARY

When the units of a given problem are written as a complex fraction, one simple rule will help convert this more difficult problem to a basic one:

- **Invert the denominator and multiply**

For example:

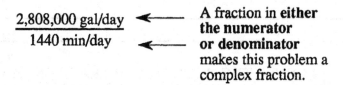

$$\frac{2{,}808{,}000 \text{ gal/day}}{1440 \text{ min/day}} \longleftarrow$$

A fraction in **either the numerator** ⟵
or denominator ⟵
makes this problem a complex fraction.

To convert this complex problem to a more basic problem, invert the denominator and multiply:

$$\frac{\dfrac{\text{gal}}{\text{day}}}{\dfrac{\text{min}}{\text{day}}} = \boxed{\frac{\text{gal}}{\text{day}} \cdot \frac{\text{day}}{\text{min}}}$$

The most difficult problems of dimensional analysis include units in complex fractions. A complex fraction has a fraction in the numerator or denominator or both. For **complex fractions** that involve units rather than numbers, the two common forms are shown in the box to the right.

ONE STEP CONVERTS A COMPLEX PROBLEM INTO A SIMPLE ONE

When using dimensional analysis for complex fractions, one simple step is required:

- **Invert the denominator and multiply.**

To invert a fraction means to flip it—the numerator becomes the denominator and the denominator becomes the numerator.

The box to the right illustrates how complex fractions may be converted to simple fractions.

COMPLEX FRACTIONS ARE FRACTIONS WITHIN FRACTIONS

Both of the following fractions are considered **complex fractions:**

$$\frac{gal}{gal/day} \longleftarrow \textit{fraction in the denominator}$$

$$\frac{cu\ ft/day}{min/day} \longleftarrow \textit{fraction in both the numerator and denominator}$$

CONVERTING THE COMPLEX FRACTIONS ABOVE TO SIMPLE FRACTIONS

First, write the fractions in vertical format, as done in any dimensional analysis problem, then **invert the denominator and multiply:**

$$\cfrac{gal}{\cfrac{gal}{day}} \longrightarrow \cancel{gal} \cdot \frac{day}{\cancel{gal}} = day$$

$$\cfrac{\cfrac{cu\ ft}{day}}{\cfrac{min}{day}} \longrightarrow \frac{cu\ ft}{\cancel{day}} \cdot \frac{\cancel{day}}{min} = \frac{cu\ ft}{day}$$

Example 1: (Dimensional Analysis—Complex)
❑ Using dimensional analysis, determine the units of the answer given the math set-up shown below.

$$\frac{756 \text{ cu ft}}{27 \text{ cu ft/cu yd}}$$

First, write the fraction of the denominator in vertical format, then divide out terms if possible. In this problem the numbers will be included as well:

$$\frac{756 \text{ cu ft}}{\dfrac{27 \text{ cu ft}}{1 \text{ cu yd}}} = (756 \;\cancel{\text{cu ft}})\frac{(1 \text{ cu yd})}{27 \;\cancel{\text{cu ft}}}$$

$$= \frac{756 \text{ cu yd}}{27}$$

The units of the answer are cu yd. The numerical answer can be completed at this point.

In Examples 1 and 2, the numerical part of the problem is included in the analysis as well. This was done to illustrate how a problem might be calculated in a dimensional analysis type of format. When checking units, however, **it is common to examine only the units** since this method is intended simply to verify math set-up and should be completed quickly.

Example 2: (Dimensional Analysis—Complex)
❑ Determine the units of the answer for the problem shown below, using dimensional analysis.

$$\frac{275 \text{ gpm}}{(7.48 \text{ gal/cu ft})(60 \text{ sec/min})}$$

First, rewrite the abbreviated term (gpm) and write the fractions in vertical format:

$$\frac{\dfrac{275 \text{ gal}}{1 \text{ min}}}{\dfrac{7.48 \text{ gal}}{1 \text{ cu ft}} \cdot \dfrac{60 \text{ sec}}{1 \text{ min}}}$$

Then invert both denominators and multiply:

$$\frac{275 \;\cancel{\text{gal}}}{1 \;\cancel{\text{min}}} \cdot \frac{1 \text{ cu ft}}{7.48 \;\cancel{\text{gal}}} \cdot \frac{1 \;\cancel{\text{min}}}{60 \text{ sec}} = \frac{275}{(7.48)(60)} \frac{\text{cu ft}}{\text{sec}}$$

The units of the answer are cu ft/sec. (The numerical part of the problem can be calculated at this point.)

PRACTICE PROBLEMS 15.2: Dimensional Analysis And Complex Fractions

❑ Use dimensional analysis to determine the units of the answer for each problem shown below.

1. (4140 gpm)(60 sec/min) =

ANS _____

2. $\dfrac{(8800 \text{ cu ft})(1440 \text{ min/day})}{6.2 \text{ cu ft/day}}$ =

ANS _____

3. $\dfrac{587 \text{ gal}}{246 \text{ gph}}$ =

ANS _____

❑ The mathematical set-up for two problems is given below. Do the units of the problem result in the desired units of the answer? If not, what should the set-up be?

4. $\dfrac{(40 \text{ in.})(1.5 \text{ ft})(2.3 \text{ fps})}{12 \text{ in./ft}}$ = cfm

ANS _____

5. $\dfrac{\dfrac{2{,}400{,}000 \text{ gpd}}{7.48 \text{ gal/cu ft}}}{635{,}400 \text{ sq ft}}$ = ft/day

ANS _____

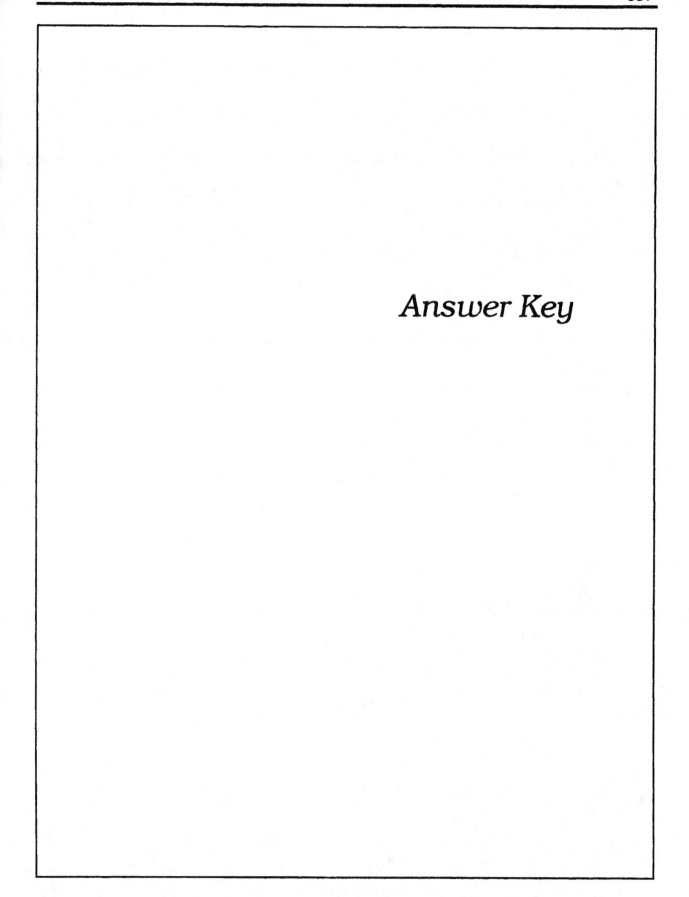

Answer Key

Basic Math Concepts Answer Key

Chapter 2

Skills Check—Chapter 2

Section 2.1 **1.** 1.8 **2.** 5.7 **3.** 5.3 **4.** 5,976,990 **5.** 8256.6 **6.** 8.06 **7.** 0.005 **8.** 360 **9.** 1649.4 **10.** 244.7

Section 2.2 **1.** 10 **2.** 90 **3.** 80 **4.** 60 **5.** 80

Section 2.3 **1.** 8 **2.** 0.8 **3.** 72 **4.** 0.6 **5.** 26

Section 2.4 **1.** 75 **2.** 5.2 **3.** 41,310 **4.** 0.038 **5.** 0.02

Practice Problems—Chapter 2

Section 2.1 **1.** 11.0 **2.** 5.0 **3.** 7993.9 **4.** 590.4 **5.** 2816.7 **6.** 4903.5 **7.** 547,616 **8.** 117.3 **9.** 508,000 **10.** 0.35

Section 2.2 **1.** 80 **2.** 12 **3.** 40 **4.** 0.83 **5.** 10.9

Section 2.3 **1.** 5.8 **2.** 45 **3.** 1.9 **4.** 14 **5.** 23

Section 2.4 **1.** 104.5 **2.** 10 **3.** 29,146 **4.** 975.2 **5.** 5 **6.** 38.8 **7.** 0.019 **8.** 38,505 **9.** 8 **10.** 563.7

Chapter 3

Skills Check—Chapter 3

Section 3.1 **1.** 5 **2.** 6 **3.** 3 **4.** d **5.** a **6.** 1/6 **7.** 3/8 **8.** 5/12 **9.** 4/8 **10.** 4/9

Section 3.2 **1.** 2/10; 1/5 **2.** Multiply the numerator and denominator by the same number. Any number may be selected. For example: 3/5 x 2/2 = 6/10 **3.** 1/7 x 3/3 = 3/21 **4.** 9/11 x 2/2 = 18/22 **5.** Divide the numerator and denominator by the same number. For example: 10/18 ÷ 2/2 = 5/9 **6.** 6/36 ÷ 6/6 = 1/6 **7.** 16/56 ÷ 8/8 = 2/7 **8.** Yes, 288 **9.** No **10.** Yes, 2660

Section 3.3 **1.** 3/4 **2.** 4/5 **3.** 3/4 **4.** 3/5 **5.** 5/6 **6.** 7/19 **7.** 8/9 **8.** 3/8 **9.** 4/13 **10.** 17/30

Section 3.4 **1.** 10/15; 12/15 **2.** 15/24; 14/24 **3.** 2/12; 9/12 **4.** 5/40; 32/40 **5.** 8/36; 3/36 **6.** 8/80; 43/80 **7.** 5/20; 12/20; 10/20 **8.** 8/12; 9/12; 2/12 **9.** 14/20; 10/20; 15/20 **10.** 16/24; 21/24; 20/24

Section 3.5 **1.** 5-1/4 **2.** 1-1/2 **3.** 55/8 **4.** 86/7 **5.** 28/5 **6.** 80/3 **7.** 2-4/5 **8.** 3-3/8 **9.** 6-6/7 **10.** 1-4/5

Section 3.6 **1.** 37/40 **2.** 2-3/8 **3.** 1-9/28 **4.** 23-2/5 **5.** 3-5/36 **6.** 3/40 **7.** 11/12 **8.** 4-11/15 **9.** 17/35 **10.** 11/24

Section 3.7 **1.** 3/56 **2.** 35/54 **3.** 1-1/8 **4.** 5-2/3 **5.** 2/5 **6.** 1/6 **7.** 8-1/3 **8.** 17-1/3 **9.** 28,000 cu ft **10.** 7/20 MG

Section 3.8 **1.** 3/2 or 1-1/2 **2.** 27/28 **3.** 12 **4.** 1-2/3 **5.** 1-17/48 **6.** 3-3/4 **7.** 2-17/30 **8.** 12 **9.** 22-2/9 **10.** 1000

Section 3.9 **1.** 335/2 or 167-1/2 **2.** 27/140 **3.** 2/35 **4.** 18 **5.** 27/2 or 13-1/2

Practice Problems—Chapter 3

Section 3.1 **1.** 2 **2.** 12 **3.** 8 **4.** 27 **5.** c **6.** a **7.** d **8.** 2/7 **9.** 6/9 or 2/3 **10.** 9/25

Section 3.2 **1.** 6/12 and 1/2 **2.** 4/8 and 1/2 **3.** Multiply the numerator and denominator by the same number. For example 5/7 x 2/2 = 10/14 and 5/7 x 3/3 = 15/21 **4.** Divide the numerator and denominator using the same number. For example 132/231 ÷ 3/3 = 44/77 and 132/231 ÷ 11/11 = 12/21 **5.** Yes, 48 **6.** No **7.** Yes, 750 **8.** Yes, 135

Section 3.3 **1.** 3/4 **2.** 5/11 **3.** 2/5 **4.** 1/3 **5.** 3/5 **6.** 9/28 **7.** 3/10 **8.** 3/8 **9.** 3/4 **10.** 8/21

Section 3.4 **1.** 18 **2.** 80 **3.** 24 **4.** 60 **5.** 60 **6.** 12/24; 9/24; 14/24 **7.** 24/36; 16/36; 12/36 **8.** 12/30; 25/30; 18/30 **9.** 20/48; 9/48; 42/48 **10.** 4/6; 3/6; 5/6

Section 3.5 **1.** 4-1/3 **2.** 1-7/8 **3.** 21/8 **4.** 24/5 **5.** 28/9 **6.** 67/4 **7.** 1-11/14 **8.** 8-10/11 **9.** 2-3/8 **10.** 3-1/6

Section 3.6 **1.** 1-55/72 **2.** 2-1/12 **3.** 9-7/12 **4.** 1-1/6 **5.** 1-1/12 **6.** 7-1/56 **7.** 1/3 **8.** 89/91 **9.** 1-3/26 **10.** 35-11/21

Section 3.7 **1.** 2/9 **2.** 11/56 **3.** 15/2 or 7-1/2 **4.** 5/4 or 1-1/4 **5.** 15 **6.** 5/14 **7.** 61 **8.** 375/8 or 46-7/8 **9.** 3/2 or 1-1/2 **10.** 2/9

Section 3.8 **1.** 6/25 **2.** 36/101 **3.** 3/2 or 1-1/2 **4.** 10/3 or 3-1/3 **5.** 59/35 or 1-24/35 **6.** 13/155 **7.** 18/5 or 3-3/5 **8.** 1/2 **9.** 3/8 **10.** 512/135 or 3-107/135

Section 3.9 **1.** 145/104 or 1-41/104 **2.** 189/116 or 1-73/116 **3.** 40/7 or 5-5/7 **4.** 1/5 **5.** 57/34 or 1-23/34

Chapter 4

Skills Check—Chapter 4

Section 4.1 **1.** 7/10; 0.7 **2.** 3/10; 0.3 **3.** 1/7; 0.14 **4.** 3/10; 0.3 **5.** 1/6; 0.17 **6** .3-1/4; 3.25 **7.** 4-1/2; 4.5 **8.** 5; 5.0 **9.** 4-1/2; 4.5 **10.** 1-3/10; 1.3

Section 4.2 **1.** 62.3 **2.** 1.29 **3.** 93.24 **4.** 22.375 **5.** 396.45 **6.** 17.15 **7.** 0.707 **8.** 107.45 **9.** 25.05 **10.** 0.0108

Section 4.3 **1.** 1828.068 **2.** 0.00045 **3.** 74.1873 **4.** 282.6 **5.** 90.1875 **6.** $2067.75 **7.** 490.625 **8.** 144.10519 **9.** 458.7 **10.** 64.68

Section 4.4 **1.** 2.135 **2.** 8.49 **3.** 30,882 **4.** 21.023 **5.** 4.56 **6.** 1747.8 **7.** 1526.139 **8.** 3.86 **9.** 94.65 **10.** 4196.64

Section 4.5 **1.** 7-5/10 **2.** 92/100 **3.** 5/1000 **4.** 12-75/100 **5.** 627-8/10 **6.** 15.33 **7.** 2.07 **8.** 2140.125 **9.** 84.857 **10.** 0.0034

Practice Problems—Chapter 4

Section 4.1 **1.** 4/10; 0.4 **2.** 6/10; 0.6 **3.** 7/100; 0.07 **4.** 34/100; 0.34 **5.** 1/10; 0.1 **6.** 5-8/10; 5.8 **7.** 2-5/10; 2.5 **8.** 3-2/100; 3.02 **9.** 2-7/10; 2.7 **10.** 1-90/100; 1.90

Section 4.2 **1.** 16.6 **2.** 3.3044 **3.** 10.375 **4.** 0.3942 **5.** 0.884 **6.** 34.43 **7.** 19.032 **8.** 24.45 **9.** 35.39 **10.** 0.0063

Section 4.3 **1.** 1709.724 **2.** $678.45 **3.** 706.5 **4.** $1.\overline{64}$* **5.** 1.4637 **6.** 80.65875 **7.** 21.125 **8.** 1.179698 **9.** 74,800 **10.** 2752.2

Section 4.4 **1.** 4.381 **2.** $0.72\overline{6}$* **3.** 1146.50 **4.** 2.08 **5.** $1931.8\overline{1}$* **6.** 2.15 **7.** 3.31 **8.** 5.68 **9.** 6.00 **10.** 11,363.64

Section 4.5 **1.** 125/1000 **2.** 78/100 **3.** 1-75/100 **4.** 53-625/1000 **5.** 195-56/100 **6.** $0.2\overline{6}$* **7.** $0.\overline{6}$ **8.** 25.75 **9.** 190.125 **10.** 2.09

Chapter 5

Skills Check—Chapter 5

Section 5.1 **1.** 2/5 **2.** 31/50 **3.** 20% **4.** 17/50 **5.** 30% **6.** 80% **7.** 17/100 **8.** 120% **9.** 14/25 **10.** 41.7%

Section 5.2 **1.** 0.19 **2.** 27% **3.** 16.8% **4.** 0.48 **5.** 1.84 **6.** 72.9% **7.** 15.36% **8.** 0.66 **9.** 5.29% **10.** 0.97

Section 5.3 **1.** 88% **2.** 811.8 **3.** 30% **4.** 828.6 **5.** 11.5% **6.** 9.46 **7.** 161.1 **8.** 1792.8 **9.** 4.2% **10.** 18,152

Practice Problems—Chapter 5

Section 5.1 **1.** 60% **2.** 4/25 **3.** 12/25 **4.** 40% **5.** 4% **6.** 3/50 **7.** 20% **8.** 3/100 **9.** 14.3% **10.** 37/50

Section 5.2 **1.** 64.3% **2.** 3.46% **3.** 102% **4.** 75.3% **5.** 5.7% **6.** 0.19 **7.** 0.87 **8.** 1.08 **9.** 0.04 **10.** 0.005

---

* The line or bar over a number indicates a repeating decimal.

Practice Problems—Cont'd

Section 5.3 **1.** 1455.2 **2.** 13.419 **3.** 4175 **4.** 25,789.473 **5.** 3.7% **6.** 189.29 **7.** 13.4% **8.** 18.75% **9.** 261.6 **10.** 72.5%

Chapter 6

Skills Check—Chapter 6

Section 6.1-Mean **1.** 147 mg/L **2.** 15.25 mg/L **3.** 1.38 **4.** 5454 **5.** 0.060
Section 6.1-Median **1.** 240 **2.** 165 **3.** 123 **4.** 250 **5.** 126
Section 6.1-Mode **1.** 220 **2.** 162 **3.** 117 **4.** 230 **5.** 133
Section 6.2 **1.** 18.5, 19, 22.5, 21, 18.5 **2.** 305, 300, 298.5, 289, 288.5 **3.** 24.5, 18.5, 20 **4.** 2516; 2547; 2590; 2616 **5.** 223;223;222;223
Section 6.3 **1.** 225 **2.** 135 **3.** 2461 **4.** 131 **5.** 23

Practice Problems—Chapter 6

Section 6.1 **1.** 147 **2.** 1.01 **3.** 2324 **4.** 4641 **5.** 44 **6.** Mean = 624 MPN/100; Median = 280 MPN/100 **7.** Mean = 186 mg/L; Median = 180 mg/L **8.** Mean = 2367 mg/L; Median = 2450 mg/L **9.** 240 MPN/100 **10.** 190 mg/L
Section 6.2 **1.** 120; 122; 126; 127; 128 **2.** 256; 263; 273; 278; 286 **3.** 21; 21; 24; 26 **4.** 150; 158; 157; 155
Section 6.3 **1.** Weighted average—139.5; Regular arithmetic mean—140.6 **2.** Weighted average 232.2; Regular arithmetic mean = 232.8

Chapter 7

Skills Check—Chapter 7

Section 7.1 **1.** No **2.** Yes, 27 **3.** No **4.** No **5.** Yes, 240 **6.** No **7.** No **8.** Yes, 540 **9.** Yes, 2310 **10.** No
Section 7.2 **1.** 22 **2.** 2 **3.** 5 **4.** 360 **5.** 24 **6.** 36 **7.** 10 **8.** 60 **9.** 12 **10.** 7
Section 7.3 **1.** $4/11 = 130/x$; $x = \$357.50$ **2.** $3/x = 950/2400$; $x = 7.6$ gal **3.** $1/x = 17/90$; $x = 5.3$ barrels **4.** $1/x = 3.78531/50$; $x = 13.2$ **5.** $3/x = 1/4.6$; $x = 13.8$ cu ft

Practice Problems—Chapter 7

Section 7.1 **1.** Yes, $0.8\overline{3}$* **2.** No **3.** Yes, $0.\overline{45}$* **4.** No **5.** Yes, 54 **6.** Yes, 798 **7.** Yes, 696 **8.** Yes, 702 **9.** No **10.** No
Section 7.2 **1.** 9 **2.** 5 **3.** 24 **4.** 9 **5.** 9 **6.** 20 **7.** 20 **8.** 56 **9.** 25 **10.** 7
Section 7.3 **1.** $1/3.7853 = x/75$; $x = 19.8$ gallons **2.** $1/3.5 = x/120$; $x = 34.3$ bags **3.** $12,000/1 = x/180$; $x = 2,160,000$ gpd **4.** $5.4/80 = x/30$; $x = 2.0$ lbs **5.** $300/1 = 1240/x$; $x = 4.1$ acres

Chapter 8

Skills Check—Chapter 8

Section 8.2 **1.** 359,040 gal **2.** 4969 cu ft (if 7.48 and 8.34 factors are used); 4968 (if 62.4 factor is used) **3.** 1,551,240 lbs **4.** 7486 lbs (if 7.48 and 8.34 factors are used); 7488 lbs (if 62.4 factor is used) **5.** 9760 gal
Section 8.3 **1.** 1122 gpm **2.** 329 cfm **3.** 1524 gpm **4.** 3.3 cfs **5.** 5.4 cfs
Section 8.4 **1.** 2.3 ft **2.** 24,816 **3.** 44.4 in **4.** 0.34 mi **5.** 86.7 yds
Section 8.5 **1.** 182.2 sq yds **2.** 209,088 sq ft **3.** 14.6 sq ft **4.** 3.1 ac **5.** 20,037.6 sq ft
Section 8.6 **1.** 264,044 gal **2.** 1134 cu ft **3.** 74.3 cu ft **4.** 76,230 cu ft **5.** 9.6 ac-ft
Section 8.7 **1.** 0.032% **2.** 12,000 mg/L **3.** 100 mg/L **4.** 0.021% **5.** 2600 mg/L

* The line or bar over a number indicates a repeating decimal.

Section 8.8 **1.** 205.2 mg/*L* **2.** 155.61 mg/*L* **3.** 11.4 gpg **4.** 6.4 gpg **5.** 242.82 mg/*L*
Section 8.9 **1.** 0.145 g **2.** 2500 mL **3.** 0.2 m **4.** 0.52 L **5.** 12,000 mg
Section 8.10 **1.** 2485.7 kg **2.** 189.25 L **3.** 60.07 cfs **4.** 23,694 m³/d **5.** 22.3 ft

Practice Problems—Chapter 8

Section 8.1 **1.** 12.6 cm **2.** 3.4 m³ **3.** 40.5 yds **4.** 165.3 lbs **5.** 8.5 oz **6.** 96.0 m
7. 24.1 hp **8.** 233.6 kg **9.** 254 in. **10.** 22.38 kW
Section 8.2 **1.** 29.92 gal **2.** 133.7 cu ft **3.** 2183.4 lbs (using 7.48 and 8.34 factors); 2184 lbs
(using 62.4 factor) **4.** 500,400 lbs **5.** 179.9 gal **6.** 25,133.7 cu ft **7.** 961.8 cu ft (using 7.48
and 8.34 factors); 961.5 cu ft (using 62.4 factor) **8.** 12,829.7 gal **9.** 1247.7 lbs (using 7.48
and 8.34 factors); 1248 lbs (using 62.4 factor) **10.** 4170 lbs
Section 8.3 **1.** 1,615.7 gpm **2.** 2,620,800 gpd **3.** 6 cfs **4.** 5972 gpm **5.** 2028 gpm
6. 4,146,912 gpd **7.** 1154 gpm **8.** 2.43 MGD **9.** 1,809,562 gpd **10.** 2507 gpm
Section 8.4 **1.** 5.7 yds **2.** 10.2 ft **3.** 8976 ft **4.** 1080 in. **5.** 13.7 yds **6.** 7.2 in. **7.** 210 ft
8. 1.7 mi **9.** 142.6 ft **10.** 660 ft
Section 8.5 **1.** 7.1 sq ft **2.** 4500 sq ft **3.** 174,240 sq ft **4.** 0.5 sq ft **5.** 0.6 ac **6.** 1296 sq in.
7. 1368 sq in. **8.** 39,204 sq ft **9.** 1.4 ac **10.** 3.1 sq ft
Section 8.6 **1.** 675 cu ft **2.** 0.9 cu ft **3.** 95,832 cu ft **4.** 0.8 cu yds **5.** 2.1 ac-ft
6. 639.3 cu yds **7.** 16.2 cu ft **8.** 5184 cu in. **9.** 14.8 cu yds **10.** 2,178,000 cu ft
Section 8.7 **1.** 0.012% **2.** 15,000 mg/*L* **3.** 250 mg/*L* **4.** 0.5% **5.** 0.68% **6.** 0.0195%
Section 8.8 **1.** 106.02 mg/*L* **2.** 8.5 gpg **3.** 15.8 gpg **4.** 2.2 mg/*L* **5.** 34.2 mg/*L* **6.** 6.4 gpg
Section 8.9 **1.** 2.4 L **2.** 0.008 kg **3.** 15,000 mL **4.** 0.0022 L **5.** 0.18 m **6.** 40,000 m
7. 0.23 g **8.** 0.25 km² **9.** 155 cm³ **10.** 30,000 g
Section 8.10 **1.** 27.6 in. **2.** 1.1 m³ **3.** 32.0 m **4.** 2152 sq ft **5.** 0.9 oz **6.** 2812 kg **7.** 567 g **8.**
0.16 gal

Chapter 9

Skills Check—Chapter 9

Section 9.1 **1.** 19.2 ft **2.** 34 ft **3.** 130 ft **4.** 13 ft **5.** 82 ft
Section 9.2 **1.** 62.8 ft **2.** 266.9 ft **3.** 188.4 ft **4.** 50 ft **5.** 40 ft

Practice Problems—Chapter 9

Section 9.1 **1.** 23 ft **2.** 370 ft **3.** 120 ft **4.** 41 ft **5.** 180 ft **6.** 374 ft **7.** 26 in. **8.** 88 in.
9. 150 ft; 25 ft; 25 ft **10.** 24 in.
Section 9.2 **1.** 24 ft **2.** 60 ft **3.** 47.1 in. **4.** 125.6 ft **5.** 157 ft **6.** 12.56 ft **7.** 120 ft

Chapter 10

Skills Check—Chapter 10

Section 10.1 **1.** 84 sq ft **2.** 4 sq ft **3.** 45 sq ft **4.** 50.2 sq ft **5.** 225 sq ft **6.** 600 sq ft
7. 25 ft **8.** 50.2 sq in. **9.** 10 ft **10.** 14.1 sq in.
Section 10.2 **1.** 250 sq ft **2.** 74 sq ft **3.** 57.1 sq ft **4.** 1.25 ft **5.** 8 ft
Section 10.3 **1.** 18.8 sq ft **2.** 1099 sq ft **3.** 78.5 sq ft **4.** 25 ft **5.** 5.0 ft

Practice Problems—Chapter 10

Section 10.1 **1.** 5024 sq ft **2.** 120 sq ft **3.** 3300 sq ft **4.** 2025 sq ft **5.** 3600 sq ft **6.** 20 sq ft
7. 90 ft **8.** 24 ft **9.** 5 ft **10.** 15 ft
Section 10.2 **1.** 575 sq ft **2.** 59.3 sq ft **3.** 56 sq ft – 12 sq ft = 44 sq ft **4.** 120 sq ft
5. 6358.5 sq ft + 4050 sq ft = 10,408.5 sq ft **6.** 1589.6 sq ft – 176.6 sq ft = 1413 sq ft
7. 14.1 ft **8.** $x = 8$ ft

Practice Problems—Cont'd

Section 10.3 **1.** 5024 sq ft **2.** 125.6 sq ft **3.** 314 sq ft **4.** 7065 sq ft **5.** 1570 sq ft **6.** 25 ft **7.** 4 ft **8.** 6 ft

Chapter 11

Skills Check—Chapter 11

Section 11.1 **1.** 800 cu ft **2.** 125,600 cu ft **3.** 381.5 cu ft **4.** 64 cu ft **5.** 20.25 cu ft **6.** 50.24 cu ft **7.** 15 ft **8.** 11 ft **9.** 2 ft **10.** 15.7 cu ft
Section 11.2 **1.** 2377.5 cu ft **2.** 31,400 cu ft **3.** 8040 cu ft **4.** 150 cu ft **5.** 3035.3 cu ft

Practice Problems—Chapter 11

Section 11.1 **1.** 12,000 cu ft **2.** 57,697.5 cu ft **3.** 75 cu ft **4.** 65.4 cu ft **5.** 2826 cu ft **6.** 390 cu ft **7.** 29,437.5 cu ft **8.** 15 ft **9.** 1800 cu ft **10.** 3532.5 cu ft
Section 11.2 **1.** 435 cu ft **2.** 79,547 cu ft **3.** 3900 cu ft **4.** 2977.5 cu ft **5.** 113 cu ft

Chapter 12

Skills Check—Chapter 12

Section 12.1 **1.** 1/5; 7/10 **2.** 2/3; 2-2/3 **3.** 2-3/8; 3-7/16
4. 15-4/5; 17-1/2 **5.** 0.07 **6.** 0.15
Section 12.2 **1.** March, April, May, June, August, December **2.** 8 in. **3.** 84 lbs **4.** Monday and Thursday **5.** 67% **6.** 0.004 ft/ft **7.** 0.08 ft/ft **8.** zone of carbonate deposits **9.** zone of Ca CO_3 solution equilibrium **10.** 7.2 to 7.6

Practice Problems—Chapter 12

Section 12.1 **1.** 1/3; 3-2/3 **2.** 7-1/2; 9-3/8 **3.** 5/16; 27/32 **4.** 7; 30 **5.** 0.00175; 0.0062
Section 12.2 **1.** 1.5 mg/L

2. 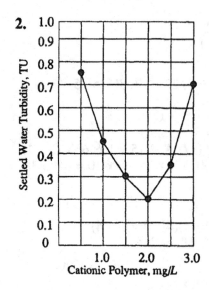   **3.** 0.45 cfs

4.

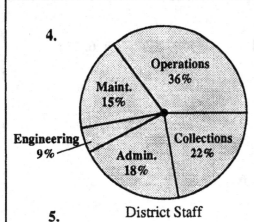

District Staff

$\frac{18}{100}$ x 360° = 65° Administration

$\frac{36}{100}$ x 360° = 130° Operations

$\frac{15}{100}$ x 360° = 54° Maintenance

$\frac{22}{100}$ x 360° = 79° Collections

$\frac{9}{100}$ x 360° = $\underline{32°}$ Engineering
$\phantom{\frac{9}{100} x 360° = }360°$

| 9% Engineering |
| 22% Collections |
| 15% Maint. |
| 36% Operations |
| 18% Admin. |

District Staff

5.

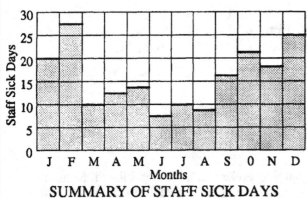

SUMMARY OF STAFF SICK DAYS

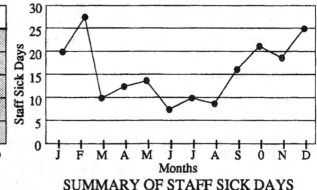

SUMMARY OF STAFF SICK DAYS

6. 8.8 lbs/day chlorine

Chapter 13

Skills Check—Chapter 13

Section 13.1 **1.** (7)(7) **2.** (12)(12)(12) **3.** 1 **4.** (3)(3)(2)(2)(2) **5.** 4^2 **6.** $16^2 3^3$ **7.** x^3
8. $(0.785)(90^2)$ **9.** 4415.6 **10.** 216

Section 13.2 **1.** $\frac{1}{(70)(70)}$ **2.** $\frac{1}{15}$ **3.** $\frac{1}{(4)(4)(4)}$ **4.** $\frac{(50)(50)}{(2)(2)(2)}$ **5.** 25^{-2} **6.** 3^{-4}

7. $(0.785)(60^2)(20^{-2})$ **8.** $28^2 \cdot 50^{-1}$ **9.** 1.125 **10.** 0.14

Section 13.3 **1.** $\sqrt{225}$ **2.** $\sqrt[3]{40}$ **3.** $\sqrt{2.5^5}$ **4.** $\sqrt{4^3}$ **5.** $1600^{1/2}$ **6.** $27^{1/3}$ **7.** $8(6358.5)^{1/2}$
8. $45^{2/2}$ or 45^1 **9.** 70 **10.** 27

Section 13.4 **1.** 2^5 **2.** $5^{-1} \cdot 3^3$ **3.** $7x$ **4.** $15^2 \cdot 2^2 \cdot 7^{-2}$ **5.** $3 \cdot 4^2 \cdot x^2$

Section 13.5 **1.** 4500 **2.** 0.675 **3.** 30 **4.** 0.001285 **5.** 2.746

Section 13.6 **1.** 1620 **2.** 0.000479 **3.** 0.09154 **4.** 6,440,000 **5.** 4.15×10^6 **6.** 3.26×10^5
7. 6.2×10^{-4} **8.** 2.5×10^3 **9.** $4.234 \times 10^9 = 4,234,000,000$ **10.** $47.6238 \times 10 = 476.238$

Practice Problems—Chapter 13

Section 13.1 **1.** (2)(2)(2)(2)(6)(6) **2.** (5)(x)(x)(x) **3.** (9)(9)(4)(4)(4)(4)(4)(4)
4. (2)(2)(2)(3)(3)(3) **5.** $2 \cdot 3^2$ **6.** 3^3 **7.** $8x^2$ **8.** 2^5 **9.** 3456 **10.** $648x^2$

Section 13.2 **1.** $\frac{6}{x^2}$ **2.** $\frac{2^4}{3^2 \cdot 7^3}$ **3.** $\frac{8^2}{5^3}$ **4.** $\frac{2^2 \cdot x^2}{9^3}$ **5.** $\frac{(2)(2)(2)}{(5)(3)(3)}$ **6.** $(9)(x)(x)(2)(2)$

7. $\frac{(3)(3)(3)(3)(1)}{(2)(2)(2)}$ **8.** $\frac{(6)(6)}{(10)(10)}$ **9.** 0.25 **10.** 5.625

Section 13.3 **1.** $\sqrt{20}$ **2.** $\sqrt[3]{1728}$ **3.** $\sqrt[3]{50^2}$ **4.** $\sqrt{2.8^5}$ **5.** $30^{3/2}$ **6.** $25^{2/3}$ **7.** $7^{3/4}$ **8.** 125 **9.** 156
10. 18

Section 13.4 **1.** $17x^2$ **2.** $4^2 5^{-1}$ **3.** $3^{-1}x^{-1}$ **4.** $3^2 x^{-3}$ **5.** $2x^{-4}y^{-6}z$ **6.** $2x^{-4}y^{-2}z^2$ **7.** 42
8. 21.3 **9.** 100 **10.** 1.7

Section 13.5 **1.** 16,300 **2.** 0.61 **3.** 822 **4.** 46,200 **5.** 0.062 **6.** 0.2409 **7.** 9900 **8.** 360,000
9. 4.26 **10.** 0.0787

Section 13.6 **1.** 1.621×10^4 **2.** 1.45×10^2 **3.** 2.8×10^{-3} **4.** 4.64×10^6 **5.** 2150 **6.** 0.0000677
7. 8,915,000 **8.** $23.1648 \times 10^5 = 2,316,480$ **9.** $1.3385074 \times 10^6 = 133,850.74$
10. $3.822 \times 10^4 = 38,220$

Chapter 14

Skills Check—Chapter 14

Section 14.1 **1.** 19.8 **2.** 0.2 **3.** 46,200 **4.** 100 **5.** 640,000
Section 14.2 **1.** 610,000 **2.** 89,000 **3.** 0.18 **4.** 0.11 **5.** 3,200,000
Section 14.3 **1.** 3000 **2.** 3600 **3.** 800 **4.** 5 **5.** 24,000

Practice Problems—Chapter 14

Section 14.1 **1.** 3760 **2.** 697,000 **3.** 4,000,000 **4.** 192.17 **5.** 40,000 **6.** 0.256 **7.** 30 **8.** 190
9. 4,700,000 **10.** 3.1746
Section 14.2 **1.** 6 **2.** 4 **3.** 4 **4.** 3 **5.** 72.0 **6.** 422 **7.** 0.0271 **8.** 49,800 **9.** 48,000 **10.** 350
Section 14.3 **1.** $(1)(40)(40) = 1600$ **2.** $(10,000)(90)(50) = 45,000,000$ **3.** $(3000)(0.01)(8) =$
240.00 **4.** $(3)(60)(1000)(7) = 1,400,000$ **5.** $(1)(60)(60)(30)(7) = 800,000$
6. $(40)(20)(8)(7) = 48,000$ **7.** $\dfrac{(4)(3)}{120} = 0.1$ **8.** $\dfrac{(20)(2\cancel{0,000})}{3\cancel{0,000}} = \dfrac{40}{3} = \dfrac{30}{3} = 10$

9. $\dfrac{2,000,\cancel{000}}{(1)(8\cancel{0})(8\cancel{0})} = \dfrac{2000\cancel{0}}{6\cancel{0}} = \dfrac{1800}{3} = 600$ **10.** $\dfrac{(10)(60)(1000)(0.2)}{(3)(3)(40)(7)} = \dfrac{12000\cancel{0}.0}{280\cancel{0}} = \dfrac{120\cancel{0}}{3\cancel{0}} = 40$

Chapter 15

Skills Check—Chapter 15

Section 15.1 **1.** cu ft/sec **2.** sq ft **3.** cu ft/min **4.** gal/min **5.** cu ft/min
Section 15.2 **1.** gal/sec **2.** ft/min **3.** days **4.** hours **5.** cu yds

Practice Problems—Chapter 15

Section 15.1 **1.** ft^3 or cu ft **2.** cu ft/day **3.** cu ft/sec **4.** No; (ft/min)(1 min/60 sec) = ft/sec **5.**
Yes
Section 15.2 **1.** $\dfrac{\text{gal}}{\text{min}} \times \dfrac{\text{sec}}{\text{min}} = \dfrac{(\text{gal})(\text{sec})}{\text{min}^2}$ **2.** $\dfrac{(\cancel{\text{cu ft}})(\text{min})(\cancel{\text{day}})}{\cancel{\text{day}}\ \cancel{\text{cu ft}}} = \text{min}$

3. $\dfrac{(\cancel{\text{gal}})(\text{hrs})}{\cancel{\text{gal}}} = \text{hrs}$ **4.** No; $\dfrac{(\cancel{\text{in.}})(\text{ft})\cancel{(\text{ft})}(\text{ft})(\cancel{\text{sec}})}{\cancel{\text{sec}}\ \cancel{\text{in.}}\ \text{min}} = \dfrac{\text{cu ft}}{\text{min}}$ **5.** Yes

NOTES:

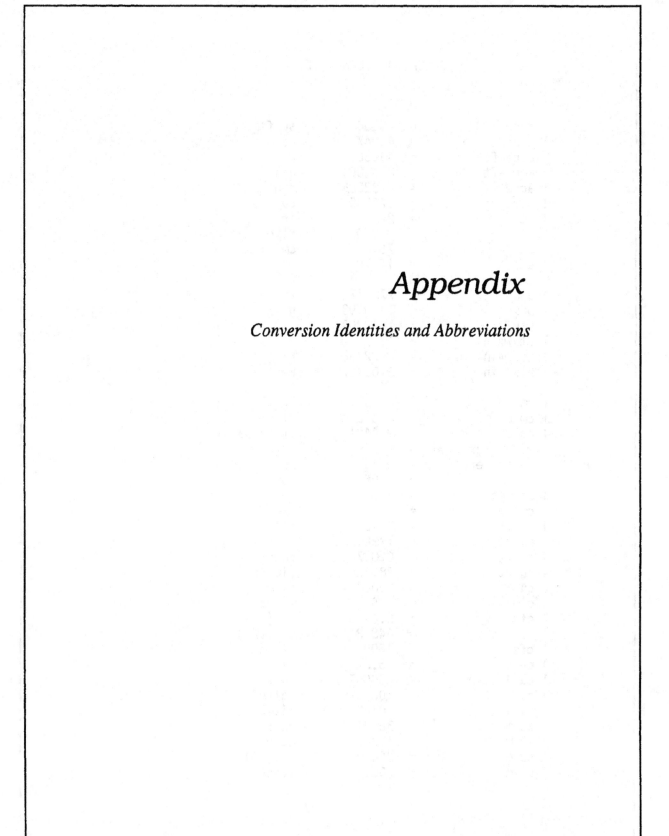

Appendix

Conversion Identities and Abbreviations

CONVERSION IDENTITIES

1 acre	=	43,560	sq ft
1 acre	=	4,047	sq m
1 acre-ft	=	43,560	cu ft
1 acre-ft	=	325,851	gal (US)
1 acre-ft	=	0.32585	mgd
1 acre-ft	=	1233.5	cu m
1 atm	=	29.92	in. of mercury
1 atm	=	33.90	ft of water
1 atm	=	76.0	cm of mercury
1 atm	=	14.70	psi
1 Btu	=	778.17	ft lb
1 Btu	=	.000393	hp-hr
1 Btu	=	.000293	kwh
1 Btu/min	=	12.96	ft lb/sec
1 Btu/min	=	0.02356	hp
1 Btu/min	=	0.01757	kw
1 cm	=	0.3937	in.
1 cm of mercury	=	0.01316	atm
1 cm of mercury	=	0.4461	ft of water
1 cm of mercury	=	27.85	psf
1 cm of mercury	=	0.1934	psi
1 cm/sec	=	1.969	fpm
1 cm/sec	=	0.0328	fps
1 cm/sec	=	0.6	m/min
1 cu ft	=	7.48	gal
1 cu ft	=	1728	cu in.
1 cu ft	=	0.0370	cu yd
1 cu ft	=	28.32	liters
1 cu ft	=	0.02832	cu m
1 cfm	=	0.472	l/sec
1 cfs	=	0.6463	mgd(US)
1 cfs	=	448.8	gpm (US)
1 cfs	=	1699	l/min
1 cu in.	=	16.39	ml
1 cu in.	=	.0005787	cu ft
1 cu in.	=	.004329	gal(US)
1 cu in.	=	.01639	liters
1 cu m	=	35.31	cu ft
1 cu m	=	1.308	cu yd
1 cu m	=	264.2	gal(US)
1 cu m	=	1000	liters

1 cu yd	=	27	cu ft
1 cu yd	=	46,656	cu in.
1 cu yd	=	0.7646	cu m
1 cu yd	=	202.0	gal (US)
1 cu yd	=	764.6	liters
1 ft	=	30.48	cm
1 ft	=	0.3048	m
1 ft of water	=	0.8826	in. of mercury
1 ft of water	=	0.4335	psi
1 ft of water	=	62.43	psf
1 ft of water	=	0.02950	atm
1 fpm	=	0.508	cm/sec
1 fps	=	30.48	cm/sec
1 fps	=	18.29	m/min
1 fps	=	1.097	km/hr
1 ft-lb	=	.001285	Btu
1 gal (US)	=	.0003068	acre-ft
1 gal (US)	=	0.1337	cu ft
1 gal (US)	=	231	cu in.
1 gal (US)	=	3785	ml
1 gal (US)	=	.003785	cu m
1 gal (US)	=	3.785	liters
1 gal (Imp)	=	1.20094	gal (US)
1 gal (US)	=	0.83267	Imperial gal
1 gal water (US)	=	8.34	lb of water
1 gpm	=	.002228	cfs
1 gpm	=	0.0631	liters/sec
1 gpm	=	8.0208	cfh
1 grain (troy)	=	0.06480	g
1 grain (troy)	=	7000	lb
1 grain/US gal	=	17.1	mg/l
1 grain/US gal	=	142.86	lb/mil gal
1 grain/Imp gal	=	14.254	mg/l
1 g	=	15.43	grains
1 g	=	0.0353	oz
1 g/l	=	58.4	grains/gal
1 g/l	=	8.345	lb/1000 gal
1 g/l	=	1000	mg/l
1 hp	=	42.44	Btu/min
1 hp	=	33,000	ft lb/min
1 hp	=	550	ft lb/sec
1 hp	=	0.746	kw
1 in.	=	2.540	cm

```
1 in. of mercury   =   1.133       ft of water
1 in. of mercury   =   0.4912      psi
1 in. of mercury   =   0.0334      atm
1 in. of water     =   0.0736      in. of mercury
1 in. of water     =   0.0361      psi

1 kg       =   2.205      lb
1 km       =   3281       ft
1 km       =   0.6214     mi
1 km/hr    =   0.9113     fps
1 km/hr    =   27.78      cm/sec
1 kw       =   56.88      Btu/min
1 kw       =   737.6      ft-lb/sec
1 kw       =   1.341      hp

1 liter    =   0.2642     gal (US)
1 liter    =   61.025     cu in.
1 liter    =   0.0353     cu ft

1 m        =   3.281      ft
1 m        =   39.37      in.
1 m        =   1.094      yd
1 mi       =   5280       ft
1 mi       =   1.609      km
1 mi/min   =   88         fps
1 mi/min   =   1.609      km/min
1 mg/l     =   1          ppm
1 mg/l     =   0.0584     grains/US gal
1 mg/l     =   0.07016    grains/Imp gal
1 mgd      =   1.55       cfs
1 mil gal  =   306.89     acre-ft
1 ml       =   .06102     cu in.
1 ml       =   .0002642   gal (US)
1 ml       =   .001       liter
1 miner's in.  =   1.5    cfm

1 oz (avdp)    =   28.3495    g
1 oz (avdp)    =   437.5      grains
1 oz (fluid)   =   1.805      cu in.
1 oz (fluid)   =   29.57      ml

1 lb           =   16         oz
1 lb           =   7000       grains
1 lb           =   453.5924   grams
1 lb of water  =   0.0160     cu ft
1 lb of water  =   27.68      cu in.
1 lb of water  =   0.1198     gal
1 lb/cu ft     =   0.0160     g/ml
```

1 lb/cu ft	=	16.02	kg/cu m
1 lb/cu ft	=	.0005787	lb/cu in.
1 lb/ft	=	1.488	kg/m
1 lb/in.	=	178.6	g/cm
1 psf	=	0.0160	ft of water
1 psf	=	4.883	kg/sq m
1 psi	=	2.307	ft of water
1 psi	=	2.036	in. of mercury
1 psi	=	0.0680	atm
1 psi	=	703.1	kg/sq m
1 sq cm	=	0.1550	sq in.
1 sq cm	=	.001076	sq ft
1 sq ft	=	144	sq in.
1 sq ft	=	0.0929	sq m
1 sq in.	=	6.452	sq cm
1 sq km	=	247	acres
1 sq m	=	10.76	sq ft
1 sq m	=	1.196	sq yd
1 sq mi	=	640	acres
1 sq mi	=	2.590	sq km
1 sq yd	=	9	sq ft
1 sq yd	=	0.8361	sq m
1 watt	=	0.056	Btu/min
1 watt	=	0.7376	ft-lb/sec
1 watt	=	.00134	hp
1 yd	=	0.9144	m

NOTES:

NOTES:

354

NOTES:

NOTES:

Printed in the United States
by Baker & Taylor Publisher Services